ENVIRONMENTAL FEDERALISM

In *Environmental Federalism*, Luke Fowler helps to refocus much-needed attention on the role of state governments in environmental policy creation and implementation in the United States. While the national government receives most of the attention when it comes to environmental policy, state governments play a vital role in protecting our natural resources. Legacy problems, like air, water, and land pollution, present one set of challenges for environmental federalism, but new problems emerging as a result of climate change further test the bounds of federal institutions.

Examining patterns of pollution and case studies from the Clean Air Act and the Clean Water Act, Fowler explores two questions: has environmental federalism worked in managing legacy environmental problems, and can it work to manage climate change? In order to answer these questions, Fowler extends James Lester's typology using political incentives and administrative capacities to identify four types of states (progressive, delayers, strugglers, and regressives) and assesses how they are linked to the success of federal environmental programs and conflicts in intergovernmental relations. He then considers what lessons we can learn from these programs and whether those lessons can help us better understand climate policy and multi-level institutions for environmental governance.

This timely read will be a valuable contribution to students, researchers, and scholars of political science, public policy, public administration, and environmental studies.

Luke Fowler is Associate Professor in Public Policy and Administration and Director of the MPA program at Boise State University, USA. His research interests include policy implementation, collaboration and collective action, and state and local government, and he has written extensively on subnational environmental policies. His related work has appeared in the *American Review of Public Administration*, *Environmental Politics*, *Governance*, *Journal of Environmental Planning & Management*, *State & Local Government Review*, *Policy Studies Journal*, *Public Administration*, *Publius*, *Public Works Management & Policy*, *Public Performance & Management Review*, and *Review of Policy Research*.

"Fowler offers a fresh take on environmental federalism by exploring how political and administrative contexts shape state environmental policies, outcomes, and intergovernmental relationships. A one-of-a kind, comprehensive analytical inquiry invaluable for students of political science, public administration and environmental policy, it is the best book written on this topic."

Richard Feiock, *Florida State University*

"The greatest challenge in environmental protection and climate change is that they require tremendous amounts of cooperation across governments. Fowler's text provides a new frame for federalism scholarship – a synergy because of environmentalism and multi-layered governance. This book is worth a read for students, scholars and anyone interested in getting a handle on this important topic."

Jessica Terman, *George Mason Schar School of Policy and Government*

"An impressive examination of the crucial – though often unappreciated – role that federalism plays in US environmental policy. Drawing on an extensive array of original data, Fowler documents significant variation in the ways that states have fulfilled their environmental protection responsibilities. His findings raise profound questions about the capability of the American system to respond to new, complex environmental challenges such as climate change."

Neal D. Woods, *University of South Carolina*

"*Environmental Federalism: Old Legacies and New Challenges* will likely become the next standard text for understanding the complex interworking of environmental policymaking in our federal system. It offers a cogent exploration of an ever changing and complex policy arena – environmental federalism. The field is in need of a contemporary update of environmental federalism and Fowler has delivered!"

Hunter Bacot, *UNC Greensboro*

ENVIRONMENTAL FEDERALISM

Old Legacies and New Challenges

Luke Fowler

Routledge
Taylor & Francis Group

NEW YORK AND LONDON

First published 2020
by Routledge
52 Vanderbilt Avenue, New York, NY 10017

and by Routledge
2 Park Square, Milton Park, Abingdon, Oxon, OX14 4RN

Routledge is an imprint of the Taylor & Francis Group, an informa business

Library of Congress Cataloging-in-Publication Data
A catalog record for this book has been requested

ISBN: 978-0-367-49096-6 (hbk)
ISBN: 978-0-367-49094-2 (pbk)
ISBN: 978-1-003-04448-2 (ebk)

Typeset in Bembo
by Apex CoVantage, LLC

To my wife, Lindsey, who always supports me in my endeavors.

CONTENTS

ILLUSTRATIONS

Tables

Figure

MAPS

ACRONYMS

Legislation:

Air Pollution Control Act of 1955	APCA
Clean Air Act of 1963 (including amendments of 1970, 1977, and 1990)	CAA
Clean Smokestacks Act of 2002	CSA
Clean Water Act of 1977	CWA
Federal Insecticide, Fungicide, and Rodenticide Act of 1972	FIFRA
Federal Water Pollution Control Act of 1948	FWPCA
National Environmental Protection Act of 1969	NEPA
Resource Conservation & Recovery Act of 1976	RCRA
Safe Drinking Water Act of 1974	SDWA

Organizations:

California Environmental Protection Agency	CEPA
Center for Climate and Energy Solutions	CCES
Database for State Incentives for Renewables and Efficiency	DSIRE
Environmental Council of the States	ECOS
Florida Department of Environmental Protection	FDEP
International Center for Technology Assessment	ICTA
Louisiana Department of Environmental Quality	LDEQ
National Institute of Money in Politics	NIMP
Non-Governmental Organization	NGO
Ohio Environmental Protection Agency	OEPA
Tennessee Valley Authority	TVA
Texas Commission on Environmental Quality	TCEQ
US Bureau of Economic Analysis	BEA

US Environmental Protection Agency	EPA
US Office of Management and Budget	OMB
West Virginia Department of Environmental Protection	WVDEP
World Resources Institute	WRI

Programs and Policies:

Federal Implementation Plan	FIP
Integrated Regional Water Management	IRWM
National Ambient Air Quality Standards	NAAQS
National Pollutant Discharge Elimination System	NPDES
Public Benefits Funds	PBF
Renewable Portfolio Standards	RPS
State Implementation Plans	SIP
State Review Framework for Compliance and Enforcement Performance	SRF
Total Maximum Daily Loads	TMDLs
Toxics Release Inventory	TRI

Technical:

Carbon Dioxide	CO_2
Consumer Price Index	CPI
Fiscal Year	FY
Greenhouse Gases	GHGs
Land Use Change and Forestry	LUCF
Methane	CH_4
Metric Tons of Carbon Dioxide Equivalent	MTCO2e
Multilevel Regression and Postestimation	MRP
Nitrous Oxide	NO
Ordinary Least Squares (regression)	OLS
Variance Inflation Factor	VIF

1

THE CHALLENGES OF SHARED SOVEREIGNTY

In 1932, the United States (US) was in the midst of a defining moment. The Great Depression had set an upstart nation on the brink of economic collapse. Failed banks and barren farms defined day-to-day lives. Herbert Hoover occupied the White House, and droves of homeless men occupied Hoovervilles across America. Franklin Roosevelt had yet to use the New Deal to transform the US economy (Kennedy, 2001; Hiltzik, 2011). Neither our policymakers nor our economic institutions were prepared for the magnitude of the challenge. Like other great crises of American history before it, the Great Depression would challenge the very foundation of the federal system. While Associate Supreme Court Justice Louis Brandeis was a lesser-known character in this story, he would coin a phrase that would carry on to encapsulate the importance of shared sovereignty for the progress of public policy: "laboratories of democracy" (New State Ice Co. v. Liebmann, 1932). But, more than that, he would also put forth an important argument for how America responds to new challenges:

> Yet the advances in the exact sciences and the achievements in invention remind us that the seemingly impossible sometimes happens. There are many men now living who were in the habit of using the age-old expression: "It is as impossible as flying." The discoveries in physical science, the triumphs in invention, attest the value of the process of trial and error. In large measure, these advances have been due to experimentation. In those fields experimentation has, for two centuries, been not only free but encouraged. Some people assert that our present plight is due, in part, to the limitations set by courts upon experimentation in the fields of social and economic science; and to the discouragement to which proposals for

betterment there have been subjected otherwise. There must be power in the states and the nation to remold, through experimentation, our economic practices and institutions to meet changing social and economic needs. I cannot believe that the framers of the Fourteenth Amendment, or the states which ratified it, intended to deprive us of the power to correct the evils of technological unemployment and excess productive capacity which have attended progress in the useful arts. To stay experimentation in things social and economic is a grave responsibility. Denial of the right to experiment may be fraught with serious consequences to the nation. It is one of the happy incidents of the federal system that a single courageous state may, if its citizens choose, serve as a laboratory; and try novel social and economic experiments without risk to the rest of the country.

(New State Ice Co. v. Liebmann, 1932)

Facing the Great Depression, America undertook a bold experiment with the New Deal and ushered in the modern economic age with new programs aimed at transforming every aspect of our economy. But the New Deal was not just a set of policies. It was a wholesale reorganization of our economic institutions, and it required Roosevelt to challenge the norms of political power both in Washington and in cities across the nation (Kennedy, 2001; Hiltzik, 2011; Crafts & Fearon, 2013). It required us to adapt to a problem that the men who founded our nation could have never anticipated, and we had to do it in spite of men who would desperately cling to the status quo. It required us to rethink not only the role of government in our society, but also how federal and state governments work together in achieving a shared vision. Like any good experiment, it taught us many important lessons about governing during trying times (Crafts & Fearon, 2013). We learned that the institutions of yesterday cannot solve the problems of tomorrow. We learned that federal and state governments cannot divvy up policy spheres between them, and instead have to find a way to cooperate. We learned that great challenges lead to great changes.

In 2019, we face a different crisis though. One that again challenges the foundations of our economy and our government: climate change. It is in many ways a problem like none other the world has ever faced. One for which previous crises which defined nations have not prepared us. It has led some to call for a Green New Deal. Yet, we have not found our "Green" Roosevelt or our way forward. However, just like the Great Depression, many are out there experimenting with new and different solutions to the technological, social, and economic challenges that underline climate change (Fiorino, 2000; Ney & Verweij, 2015). While some of the big lessons from the Great Depression may help us here, lessons from a quieter crisis may be more applicable. Often forgotten as a pivotal moment in American political history are the emerging environmental problems of the post-World War II era. These legacy environmental problems, like polluted air and water, presented an unprecedented challenge at the time (Andrews, 2006). One for which, like climate change, our political, economic, and social institutions were unprepared. From this challenge, we learned new lessons about the limitations and opportunities that federalism presents.

Three decades after the New Deal, America had taken its place as the leader of the free world, economically and politically. Unfortunately, growth and development brought new problems. This time it would be poison smog and silent springs, problems that would force us to think more deeply about how mankind interacts with the environment. It would also force us to think deeply about how we organize ourselves to not only protect human health, but also to ensure that natural resources are available for future generations to enjoy. In response, a new group of innovators helped forge how the US government would deal with environmental protection in the 1970s. Although federal and state governments had dabbled in environmental policies over the previous decades, never before had we set our sights on comprehensive strategies to protect our air, land, and water from degradation and toxic pollutants (Ruckelshaus, 1985; Andrews, 2006). While this was a new kind of problem, institutional approaches pioneered during the New Deal still lingered. Again, we had to adapt our institutions and how federal and state governments worked together.

So, here is why we study environmental federalism: it is ultimately the study of how sovereign power is wielded over the environment and how we organize collective efforts to manage environmental challenges through existing institutions. While sovereign power always lay with the American people, the US Constitution entrusted the exercise of this power to both the national and state governments. This ensured that the will of one state could not be thrust upon others without due process, and established an important set of vertical checks and balances on the national government. After all, a key fear of the framers was another tyrannical government having dominion over the American continent. But this system also created fragmented regulatory regimes where crossing state lines meant entering a whole new realm of political institutions. Although a rural nation could withstand such institutional limitations, the US quickly grew into an urbanized and industrialized behemoth where economic and social activities routinely transected jurisdictions. More pressing of a concern though is that the natural world has no respect for the arbitrary jurisdictional lines that men draw. Thus, the environmental crisis of the mid-20th century was in many ways a crisis for how power could and would be shared in the new age.

It is with this motivation that a subtitle for this book was chosen: *Old Legacies and New Challenges*. In one sense, it refers to our legacy of institutions tied to the Constitution and the framers' original intent of how federal and state governments would share power versus the new challenges that would emerge over the next two centuries and create fundamental challenges to those institutions. In another sense, it refers to our legacy of environmental problems from the mid-20th century that breathed new life into our institutions (many of which are sufficiently at bay after five decades) versus the new environmental challenges (i.e., climate change) of the 21st century with which we are currently grappling. It is here that the dynamics of environmental federalism are defined: by old and new institutions on the one hand, and by old and new environmental problems on the other. As such, in the coming pages, we seek to answer two fundamental questions: 1) have the institutions of environmental federalism worked in managing legacy environmental problems? and 2) can those institutions work to manage new, more complex environmental problems?

My argument is rather straightforward: the national-state system for protecting the environment hinges on states and their ability to adapt national programs to meet the unique needs of local communities. In theory, this allows for national environmental standards to be met in a way that satisfices local interests. In practice though, states do not always have the political will or administrative capacity to make these programs work, while other states are not satisfied with the bare minimum. In turn, this sends ripples through the federal system, with national and local governments responding to how state governments fill this intermediate policy space within the federal hierarchy. While many times this leads to power struggles with the national government, local governments have begun to quietly respond with their own policy innovations to fill gaps left by fledgling state programs. These patterns of environmental federalism set the stage for understanding how climate policies can and should be managed within the federal system, and poses new questions about the institutional limitations of American government in the face of new, complex environmental challenges. Although environmental federalism is tied to an institutional legacy that was not designed for the modern age, it still creates an anchor for mechanisms of environmental governance, as actions and inactions of national and state governments play an important role in shaping how policy actors engage with environmental problems.

Policy in a Federal System

As a consequence of these challenges, managing policy in a federal system is a complex process, where different policy actors interact in the course of providing public services. Policy actors are the various elected officials, bureaucrats, and other policy entrepreneurs who have the power to influence policy decisions at the institutional-level (i.e., state legislatures or administrative agencies). Policymaking and policy implementation are the basic building blocks of public services. It is the synergy of policy in theory and policy in practice that affects how people interact with the environment (e.g., how many toxic chemicals are released by facility X during its normal manufacturing process?). While we largely view policymaking as a function of elected officials and policy implementation as a function of bureaucrats, both elected officials and bureaucrats have influence over both policy choices and their execution through their various roles in the process (Svara, 1998, 2008). Policymaking is a political process dominated by competing value judgements concerning which environmental or social conditions are problems and how those problems should be addressed. Policymaking is heavily influenced by issue framing and policy narratives, as policy entrepreneurs try to connect problems with solutions in a way that is politically acceptable. In doing so, policy entrepreneurs take advantage of ambiguity to construct advocacy coalitions that support their preferences (Kingdon, 1995; Herweg, Zahariadis, & Zohlnhofer, 2018; Jenkins-Smith, Nohrstedt, Weible, & Ingold, 2018). As such, there is an inherent bias towards thinking about how policies directly benefit target populations so costs and benefits are clearly discernible (May, 1991; Schneider, Ingram, & deLeon, 2014).

On the other hand, policy implementation is where policy in theory is put into practice based on social, political, and technical realities. Due to ambiguous policies though, policy implementers tend to use discretion in interpreting how policies apply to real-world scenarios, which provides them substantial influence over the norms of policies in practice (Hill & Hupe, 2014; Fowler, 2019). Given this, elected officials use rulemaking, resource control, and oversight powers to constrain administrative discretion and maintain political control over the bureaucracy (Wood & Waterman, 1991; Potoski & Woods, 2001; Wood & Bohte, 2004). As a result, policy implementers tend to focus on the things that are monitored and evaluated so they may appease elected officials, and the core function of most bureaucratic agencies is policy implementation (Brehm & Gates, 1999). Ultimately, this creates an inherent focus on definable problems and assessable results, so policy actors can establish accountability and communicate what they plan to do and what they have done to stakeholders. Furthermore, the path of least resistance is typically to act incrementally in order to maintain the status quo and provide stability (Lindblom, 1959; Jones & Baumgartner, 2005; Fowler, 2019). Consequently, our political institutions encourage policy actors to identify success based on policy adoption and implementation outputs and not necessarily on environmental or social outcomes.

Since authorities are dispersed both vertically and horizontally within the federal system, national, state, and local governments tend to work within the same space to adopt and implement policies based on their own jurisdictions, capacities, and interests (Wright, 1988; Agranoff & McGuire, 2001). During the cooperative federalism era, these relationships took a top-down form, which assumed that the federal system is a single institution with the national government coordinating the activities of subnational governments. In turn, subnational governments served as compliance managers in order to ensure that rules and procedures designed at the national-level are followed at local-levels. This left state governments very little wiggle room to adapt policies, pursue innovative approaches to environmental protection, or to negotiate for program adjustments in order to better meet the needs of their jurisdictions. Additionally, this approach assumes that policy can be effectively coordinated across levels of government by policy actors at the top of a hierarchical system that is distanced from where policy is actually practiced, and it is predicated on state and local governments' willing compliance with directives (Agranoff & McGuire, 2001).

Following the advent of new federalism, donor-recipient approaches became more and more common and allowed the national government to delegate authorities to state or local governments. In turn, state and local governments used their local implementation capacities to match national programs with local technical, sociopolitical, or economic challenges. Subnational governments were treated more like partners than compliance managers and negotiated program adjustments and funding as necessary (Liebschutz, 1991; Agranoff & McGuire, 2001). These donor-recipient approaches took into account how important local actors are in making policy work in practice, such as the coping mechanisms developed by street-level bureaucrats or the efforts by administrators to construct political coalitions to legitimize their administrative decisions (Lipsky, 2010; Reed, 2014). However, this approach also leads to principal-agent problems where states are able to exploit information asymmetries to work shirk or maximize funding by misleading national policymakers about the realities of policy implementation in their jurisdiction. When goal conflicts emerge, it also gives states opportunities to act in their own best interest rather than the national governments' (Waterman & Meier, 1998). As a result, decentralization tends to lead to inequitable outcomes, as well as institutions that are vulnerable to external stress (Meier & O'Toole, 2009).

For environmental policy,

> the policy challenge underlying pollution is balancing its environmental and health costs against its economic benefits. Achieving the optimal balance between the two is easier when the pollution is narrowly concentrated in a region rather than widely dispersed across different local climates and geographies.
>
> *(Woods & Potoski, 2010, p. 723)*

Thus, on the positive side, approaching environmental protection as a donor-recipient exercise places the decisions on how to balance competing interests against each other in the hands of an intermediate government that is much closer to the people than national policy actors. In theory, this system still means that we should expect consistency in environmental conditions across the country. However, in practice, there is a large degree of variation in how states use their policy authority because they take different approaches to protecting the environment in response to the unique political, socioeconomic, and technical circumstances of their jurisdictions. In some cases, it is because states fail to meet their minimum obligations, while in other cases, it is because states want to go above and beyond standards set at the national-level. In either case, the federal system is designed in such a way that states serve an essential but flexible role that creates an important degree of variation in how environmental protection is provided.

The Environment and Institutional Barriers

Mark Twain called the Mississippi River the "body of a nation." He spent a good portion of his life on its riverboats and witnessed firsthand how it created shared identities and cultures for the people living along its banks (Twain, 1984). More recently, Minnesota Senator Amy Klobuchar announced her bid for the presidency from its banks in St. Paul, speaking of how the Mississippi runs through America's heartland, connecting communities and creating opportunities (Pioneer Press, 2019). The Mississippi River is a true American symbol of the interconnectedness of people and the environment, the connection between culture and ecosystems, and the connection between heritage and natural resources. It is also an illustration of how environmental concerns make a mockery of the institutions that we so thoughtfully crafted to deal with political problems. The mighty Mississippi serves as the state line for ten states from Canada to the Gulf of Mexico, and 22 more from Montana to New York are part of its watershed. For states like Louisiana and Mississippi, that sit along its last miles, water from nearly 40% of the country flows through their territory before entering the Gulf (US Environmental Protection Agency (EPA), 2019). Millions of people rely on the river for their livelihoods, food, and cultural artifacts. And when the river shifts course, communities are literally left high and dry.

But herein lies the complexity. When farmers in North Dakota use pesticides, toxic substances make a long journey to the Gulf of Mexico (or get caught up somewhere along the way) where they negatively impact the growth development of fish and other wildlife. In turn, people relying on fish and wildlife for their livelihoods miss out on economic opportunities or consume toxic chemicals that, even in small doses, can lead to an increased risk of a litany of diseases over a lifetime. When the Army Corps of Engineers builds dams or levees to control flooding, they must also decide which ecosystems will be destroyed by an influx of water at some indeterminate time in the future, or how to distribute the benefits of controlled waterflows to communities that rely on the river as a central part of their daily lives. Although it is a much more contentious issue in the West, states up and down the river must also decide who has property rights and access to its banks and water, impacting farm irrigation, recreational activities (e.g., fishing, hunting, boating), and economic development. Ultimately, the mighty Mississippi has no respect for either the territorial divisions created in the 19th century or the government bureaus in the 20th century, so the task of maintaining it as a viable economic, social, and cultural resource transects those arbitrary lines of distinction and challenges the institutions sitting along its banks.

So, let us assume for a moment that policymakers want to pursue environmental protection as an objective but run up against limitations in their authority. When faced with such constraints, the simple solution would be to seek out others who have the said capacities and work together to solve problems. However, in the federal system, fragmented regulatory regimes create institutional barriers to doing so, which isolates our policy actors. That is, no single agency has the geographic or institutional authority to manage the entire sources or impacts of an environmental problem, so there is a piecemeal approach to dealing with interdependent pieces of a larger puzzle (Feiock, 2013). For instance, in managing fish habitats on the Snake River, state regulators in Wyoming, Idaho, and Oregon manage parts of the river separately, while federal agencies with major land holdings along the river, like the Bureau of Land Management or National Park Service, also manage their own river segments. Within those states, there are also separate agencies concerned with the river in as far as it effects general environmental quality, public health, wildlife, agriculture, transportation, and so on.

Consequently, one of two scenarios emerges: either agencies manage river segments separately, as if they are not part of a whole, or they cooperate to manage the river through coordinated efforts. Certainly, the latter sounds like a better idea when one considers the size and scope of most environmental challenges. So, why pick the former? In two words: transaction costs. Or, more precisely, the marginal utility of cooperating may be counterbalanced by the costs of doing so. In general, transaction costs emerge from the complexities of coordinating two separate organizations through a joint effort and are primarily a result of uncertainty (Moe, 1984; Brown & Potoski, 2003; Carr, LeRoux, & Shrestha, 2009). When organizations operate within their defined geographic or institutional jurisdictions, they largely have control over what happens, and, through the use of formalized rules and procedures, they create stability and predictability. However, when organizations work outside of those defined jurisdictions, they have less control, as their formal rules and procedures do little to dictate how other policy actors operate. Thus, working across organizations is a break from the status quo of established parameters and processes, which causes uncertainty in what could happen. It also requires a large investment of time, energy, and resources, so there is a risk that if cooperation fails, it is a waste of limited resources (McGuire, 2006; Feiock, Lee, Park, & Lee, 2010; Fowler, 2019).

In essence, by creating institutions and organizations, we are drawing lines around problems so that people can focus on narrow issues within an otherwise chaotic, ambiguous world (March & Olsen, 2010). However, this also causes people to concern themselves only with problems that fit within their domains. Institutions of federalism do this in two important ways that environmental problems challenge. First and foremost, states create a narrow geographic focus. This makes both political and administrative sense when one considers the broad diversity of public beliefs about government and the challenges of managing a nation with 325 million people and nearly 4 million square miles of territory. But state lines were not purposely drawn around ecosystems, so rarely does a single state have authority over the totality of an environmental problem (Stein, 2009). Consequently, there are few incentives for one state to become concerned about the environmental problems of another state. Why should water quality regulators in Georgia care about pollution in the Suwannee River if it is on the other side of the state line? Why should air quality regulators in New Jersey care about air pollution once it drifts into New York? Although there is a moral argument as to why, the mismatch between geographic jurisdictions and watersheds, airsheds, and ecosystems means that political incentives rarely encourage people to think holistically about environmental problems.

Second, we create organizations to achieve missions. Again, this makes administrative sense when one considers the depth, diversity, and complexity of activities with which government is charged (Wilson, 1989; Waldo, 2007). When new activities or problems pop up, we then consider if they fit within an existing mission or if we need to create a new mission to deal with it. After two centuries of this, there are more than 400 federal departments, agencies, and subagencies and countless more at the subnational-level (Crews, 2015). Of course, this specialization of tasks makes it easier to build capacity, develop expertise, and find efficient, effective ways to solve problems. However, it also limits the scope of how organizations see their role in society (Wilson, 1989; Boin, Kuipers, & Steenbergen, 2010). For instance, a pollution control agency may regulate stationary sources of airborne pollutant emissions; while another deals with transportation planning to reduce emissions from mobile sources; while another deals with forestry practices and smart growth planning to ensure there is sufficient tree canopy to absorb and offset airborne pollutants; while another deals with the public health issues related to human exposure. Organizational lines that separate these agencies create a barrier to the coordination of their activities (Feiock, 2013). They also create a barrier to thinking beyond stationary sources, or mobile sources, or biodiversity, or public health, and to thinking holistically about how air quality is managed in a way that satisfies competing political, social, economic, technical, and environmental demands.

In many ways, this type of segmentation of environmental problems contributes to a tragedy of the commons, which numerous social scientists have dedicate their lives to understanding (including Elinor Ostrom, who won the Nobel Prize in Economics for her work on this issue) (Ostrom, 2015). Garrett Hardin's (1968) description of how herdsmen sharing a common-use pasture may be the most famous. Hardin argued that the utility of adding an additional animal to the pasture (i.e., production of foodstuffs, profit from sale) would be fully enjoyed by the herdsmen who owned the animal, while the impact of that animal on the pasture (i.e., overgrazing, waste production) would be shared among all of the herdsmen using the pasture. Thus, there is no incentive for the individual to limit the number of animals in his herd, even though this behavior is likely to lead to overuse and eventually a pasture that cannot sustain a herd. In fact, savvy herdsmen may actually respond to this scenario by placing as many animals in the pasture as possible in order to extract its resources before it becomes overgrazed, and, as a result, speed up the pasture's deterioration.

What Hardin is describing is fairly simple. People respond to incentives that are structured by the institutions in which they function (in Hardin's case, the common pasture). Therefore, we can only expect people to act in a self-rational way when the benefits of doing so outweigh the costs of acting with the best interests of the community in mind. From this perspective, politicians then may ask: "what is the utility of allowing a few more pollutants into the air or water if the economic benefits of doing so are concentrated in my state but the negative externalities are shared with my neighbors?" Alternatively, bureau managers may ask: "what is the utility of working more diligently to improve this environmental problem if the benefits of doing so are shared but the costs are solely mine?" While scholars have begun to unravel how to overcome these types of institutional barriers, they are endemic to our federal institutions, which leads to unjust and inequitable distributions of the costs and benefits of environmental policies (Bowen & Wells, 2002; Feiock & Scholz, 2009; Ostrom, 2015).

Nevertheless, many scholars argue that people are not simply motivated by economic rationales, and the emotional and social connections shared between individuals create communities where people will deny self-interest in order to serve a larger whole. But there still is a basic cost-benefit calculus to these decisions that we must keep in mind: how do the extrinsic rewards associated with acting self-rationally balance against the intrinsic rewards of being a part of a community? Or, do we care enough about our communities to deny our own self-interest when it means a better outcome for everyone? To this end, political rewards come in different forms and are largely driven by how we value the environment and how we define our communities (Stone, 2011). Certainly, there are many who place undisturbed landscapes and pristine waterways over high-paying jobs and the luxuries of the modern world, and vice versa (Dunlap, 2008; Milfont & Duckitt, 2010). Our institutions are central to how those rewards are structured and encourage or discourage people to act in a way that is environmentally responsible, sustainable, and equitable.

Even if these political rewards exist, in order to protect the environment, our institutions must also have the requisite administrative capacities to accomplish this goal. Certainly, we cannot expect a single herdsman to safeguard the equitable and sustainable use of a common pasture on his own, nor should he if it is a common-use pasture shared among the community. Rather, it would require multiple people working together. They would need to establish rules and processes for making decisions about which behaviors would be acceptable or how benefits would be distributed among herdsmen. How else are various herdsmen to know what they should or should not do with the pasture? They would also need to develop mechanisms to encourage (or coerce) compliance, because what good are rules if no one follows them? As this system grows more complex over time and includes more people and more rules, they would need to find more efficient ways to organize themselves. Otherwise, they may find themselves bogged down by simply trying to make sense of the system they created. Collectively, these represent different types of capacities that are necessary to accomplish the greater goal of managing a common-use pasture; of course, in the absence of rules, compliance, and enforcement, we are likely to find that herdsmen continue to function self-rationally, as there are no institutions to encourage behavioral changes (Ingraham, Joyce, & Donahue, 2003; Christensen & Gazley, 2008).

The Wickedness of Climate Change

While the majority of Americans now believe that climate patterns are changing and that mankind has contributed to that phenomenon (i.e., anthropogenic), there was a time when we did not accept that climate change was real. Then, we did not accept that we could do anything about it. Now we are struggling with what to do about it (Otto, Frame, Otto, & Allen, 2015; Shwom et al., 2015). But many saw this coming. The first international summit that began a global conversation about climate change, bringing it to the forefront of international politics, was held in 1992. The United Nations Conference on Environment and Development, commonly known as the Rio Summit, brought together the international community to discuss issues of sustainability, alternative energy sources, and other mechanisms to protect the environment for peoples around the globe. Most importantly, the Rio Summit would set the stage for a comprehensive agreement to be introduced at the United Nations Framework Convention on Climate Change in Kyoto, Japan, in 1997 (i.e., the Kyoto Protocols) (Sweet, 2016). Although the US was not a party to the Kyoto Protocols, the Clinton administration was in favor of thinking globally and long-term about the consequences of how humans interact with the environment.

In Kyoto, Al Gore would sum up the political challenge as such:

> Our fundamental challenge is now to find out whether and how we can change the behaviors that are causing the problem. To do so requires humility, because the spiritual roots of our crisis are pridefulness and a

> failure to understand and respect our connections to God's Earth and to each other.
>
> *(Clinton White House Archives, 2019)*

He would also predict increasing floods and more powerful storms, crop failures and the spread of diseases, and countless economic and social consequences that would result from changing climate patterns. Sadly, Gore's predictions would largely turn out to be true. The environmental effects of climate change would only get worse over the next two decades, which in turn would take their economic and cultural tolls on communities around the world (Letcher, 2016; Hsiang et al., 2017). As a result, the world is now pursuing climate change action; or at least there are a lot of serious people trying to find a way forward. While only 84 nations signed onto the Kyoto Protocols, 185 signed onto the Paris Agreement (its successor agreement) in 2015 (Sweet, 2016). Further, when President Donald Trump announced the US would withdraw from the agreement in 2017, 22 states formed a climate alliance to continue to implement the Paris Agreement domestically, and hundreds of businesses and local governments pledged to continue to support its goals (Watts, 2017; Konisky & Woods, 2018).

Of course, climate change is by no means a new environmental problem and has had the attention of some advocates, policymakers, and scholars for decades (Bolin, 2007; Gupta, 2010). The intention here is not to frame climate change as a new *environmental problem*, but rather as a new *environmental policy problem* in that it has supplanted a legacy of environmental problems on the public radar. In other words, an environmental problem is a situation in which the natural systems are negatively impacted by human activity and vice versa, while environmental policy problems are the difficulties associated with developing equitable, effective mechanisms for collective actions to solve environmental problems. Thus, while climate change may be a long-standing environmental problem in so as far as it impacts on the natural world, it is just now becoming an environmental policy problem in that our society is taking it seriously as a problem seeking a solution. Furthermore, while legacy environmental problems are still part of the environmental policy agenda, it is difficult to discuss air or water quality without at least framing them as part of the wider conversation on climate change. This is a relatively new phenomenon in the political ordering of environmental issues, in which climate change is dominant and other environmental problems have been relegated to secondary status.

But, if so many are working to solve this problem, why is climate change so hard to deal with? Well, the simple answer is it is a wicked problem that is ill-defined, misunderstood, and too complex to solve without comprehensive involvement from an array of policy actors, economic interests, and political institutions (Rittel & Webber, 1973; Lazarus, 2009; Grundmann, 2016). The long answer is more complicated. First, there are many lingering questions and a lot of uncertainty (Dessai & Hulme, 2004; Webster et al., 2003). For instance, mitigation advocates argue that if we stymie the production of greenhouse gases (GHGs), the process can be slowed or stopped. Others argue that it is a practical impossibility to return GHGs to preindustrial revolution levels, so the only real strategy is to adapt to the changes in our ecosystem. Most reasonable minds understand that some combination of adaptation and mitigation is necessary, but this creates another complicated dimension for both managing the political debate and building the necessary institutional capacities (Kane & Shogren, 2000; King, 2004; Urwin & Jordan, 2008). There are also many unreasonable minds that believe we should do nothing at all.

Second, the causes of climate change occur at a global-level, where the aggregate GHG production from economic processes around the world create a cumulative problem. But the impacts are local (Wilbanks & Kates, 1999). This leads to a distinct separation between those who cause the problem and those who deal with the consequences, so there is little incentive for those who contribute to climate change to change their behavior, unless they are also being impacted by rising sea levels, droughts, floods, wildfires, and so on (Scannell & Gifford, 2011). Unfortunately, one of the cruel ironies of climate change is that the causes tend to be concentrated in economically developed, politically prosperous societies, while impacts are most likely to be felt by the most vulnerable populations (Wilson, Richard, Joseph, & Williams, 2010; Althor, Watson, & Fuller, 2016). To make matters worse, climate change is a slow-burn problem, so the real impacts will be spread across generations (Levin, Cashore, Bernstein, & Auld, 2012). In other words, there is essentially a mismatch between the benefits of acting irresponsibly towards our environment and the costs that climate change creates.

Third, there is a volunteer's dilemma at work. We can all make a small sacrifice that benefits everyone or wait to see if we can free-ride off of someone else's sacrifice (Lake, 1996; Vasi & Macy, 2003). At the national-level, political leaders could adopt new regulations to mitigate the contributions of climate change emissions coming from their jurisdictions. But, if they do, the costs will fall on their constituents, while the benefits will be shared by all. Consequently, the economic costs of climate action will always outweigh the benefits at disaggregate-levels (e.g., national, regional, state, local) even though they may balance out at the global-level. Adding to the difficulty, climate change is an international matter that one nation cannot solve by itself, so the costs of those new regulations may be for nothing if others do not also follow suit (Adger, 2001; Hoffmann, 2011). In essence, incentives for climate action only exist if there are political rewards associated with the sacrifice. In other words, people must find more utility with the intrinsic value of acting altruistically on climate change than the economic costs associated with doing so in order for the cost-benefit analysis to tip in the favor of action, rather than inaction.

Finally, climate change is not one problem; it is an enhancer of every problem (Letcher, 2016). Wildfires in the West have occurred for thousands of years, but in the last few decades, they are becoming more frequent and more devastating (Abatzoglou & Williams, 2016). The same is true for hurricanes in the Southeast (Pielke et al., 2005), droughts in the Southwest (MacDonald, 2010), and toxic algae blooms in coastal Florida (Paerl & Paul, 2012). Of course, these also exacerbate economic, social, and public health problems for communities around the world (Hsiang et al., 2017). The scientific challenge is understanding which portion of these disasters is caused by climate change and how we can expect these problems to evolve. The political challenge is that people overlook climate change as a factor at all and instead decide that these disasters have always occurred, so there is no reason to act (Poortinga et al., 2011; Weber & Stern, 2011).

Plan of the Book

In the modern era, states have enough discretion to chart their own path in environmental protection, so, in considering the role of states further, we need to examine the specific context in which they operate. As this chapter has laid out the basics of an argument for why environmental federalism is an important topic of research, the plan of this book is to further examine how states at the intermediate-level of government dictate the efficacy of this system. Specifically, we will examine how the political and administrative contexts in which states provide environmental protection, and how variations in those dimensions, impact the environment as well as intergovernmental relationships. We concentrate on legacy environmental policies (i.e., Clean Air Act (CAA), Clean Water Act (CWA), Resource Conservation and Recovery Act (RCRA)), as they provide us with particular insights into the consequences of state implementation for federal programs. These programs in particular have been refined over decades and should reflect the normal operations for this system. The overall goal is to then extrapolate lessons that may further our understanding of the challenges inherent in managing climate policy within federal systems.

We begin by assuming at a base-level that states want to provide environmental protection at the highest level possible within practical limitations. But how states identify the highest level possible and the practical limitations of providing public services varies between jurisdictions. For instance, some may argue that following a toxic waste spill the environment should be returned to its pre-spill conditions with zero traces of toxic chemicals, while others may argue that toxic chemicals should be remediated only to a point that they are nonhazardous to humans. The decisions concerning this are twofold. The first part is: what level of environmental remediation should be provided? This is a political choice that is subjective. It is both ethical and reasonable to choose either level based on the existing cultural norms and environmental beliefs. Thus, the choice is dependent on how policymakers balance competing political demands. The second part is: what are the practical limitations to providing that level of service? This is an administrative question that concerns authorities and capacities, or the ability to protect the environment in a practical way. Differences in how state leaders come to determine the answers to these questions should manifest in two ways: 1) outputs from environmental programs (i.e., pollution); and 2) the nature of state relationships with national and local governments.

To this end, the book will proceed as follows. In Chapter 2, we will provide a brief history of environmental federalism starting with the development of federal institutions through the 19th and early 20th centuries, followed by the establishment of comprehensive environmental protection strategies in the mid-20th century and the evolution of federal-state relationships through the late 20th and early 21st century. In Chapter 3, we will take a more in-depth look at political incentives, namely how public attitudes towards the environment, environmental interest groups, and horizontal competition create different contexts for state decision-making. In Chapter 4, we will examine the administrative capacity dimension, and how the operational abilities of state environmental agencies and policymaking institutions affect environmental conditions in terms of policymaking, information management, and creating accountability. We will also present data on how these factors vary across states, and construct political incentives and administrative capacities indices in order to make inter-state comparisons.

In Chapter 5, we will use these indices to construct a typology of states based on their political commitment to and capacities to provide environmental protection, and then examine how this typology explains patterns of pollution. In Chapter 6, we will examine the uneasy partnership for clean air between national and state governments that can be both cooperative at times and full of tension and conflict at other times. In Chapter 7, we will examine the rise of local governments in recent years, and how state action and inaction shape the role they play in environmental governance. In Chapter 8, we will consider how much of this evidence from legacy environmental programs applies to climate change, and what it tells us about the prospects for a national climate policy. Finally, in Chapter 9, we will consider the insights this examination has provided, draw conclusions about the prospects for environmental federalism, and suggest a few lingering questions that we were unable to answer here.

References

Abatzoglou, J.T. & A.P. Williams. 2016. Impact of Anthropogenic Climate Change on Wildfire across Western U.S. Forests. *Proceedings of National Academy of Sciences* 113(42): 11770–11775.

Adger, W.N. 2001. Scales of Governance and Environmental Justice for Adaptation and Mitigation of Climate Change. *Journal of International Development* 13(7): 921–931.

Agranoff, R. & M. McGuire. 2001. American Federalism and the Search for Models of Management. *Public Administration Review* 61(6): 671–681.

Althor, G., J.E.M. Watson, & R.A. Fuller. 2016. Global Mismatch between Greenhouse Gas Emissions and the Burden of Climate Change. *Scientific Reports* 6: 20281.

Andrews, R.N.L. 2006. *Managing the Environment, Managing Ourselves: A History of American Environmental Policy*, 2nd ed. New Haven, CT: Yale University Press.

Boin, A., S. Kuipers, & M. Steenbergen. 2010. The Life and Death of Public Organizations: A Question of Institutional Design? *Governance* 23(3): 385–410.

Bolin, B. 2007. *A History of the Science and Politics of Climate Change: The Role of the Intergovernmental Panel on Climate Change.* Cambridge: Cambridge University Press.

Bowen, W.M. & M.V. Wells. 2002. The Politics and Reality of Environmental Justice: A History of Considerations for Public Administrators and Policy Makers. *Public Administration Review* 62(6): 688–698.

Brehm, J. & S. Gates. 1999. *Working, Shirking, and Sabotage.* Ann Arbor, MI: University of Michigan.

Brown, T.L. & M. Potoski. 2003. Transaction Costs & Institutional Explanations for Government Service Production. *Journal of Public Administration Research & Theory* 13(4): 441–468.

Carr, J.B., K. LeRoux, & M. Shrestha. 2009. Institutional Ties, Transaction Costs, and External Service Production. *Urban Affairs Review* 44(3): 403–427.

Christensen, R.K. & B. Gazley. 2008. Capacity for Public Administration: Analysis of Meaning and Measurement. *Public Administration & Development* 28(4): 265–279.

Clinton White House Archives. 2019. *Remarks as Prepared for Delivery for Vice President Al Gore Kyoto Climate Change Conference* [online]. Available at https://clintonwhitehouse2. archives.gov/WH/EOP/OVP/speeches/kyotofin.html

Crafts, N. & P. Fearon (editors) 2013. *The Great Depression of the 1930s: Lesson for Today.* New York: Oxford University Press.

Crews, C.W. 2015. *Nobody Knows How Many Federal Agencies Exist: Competitive Enterprise Institution* [online]. Available at https://cei.org/blog/nobody-knows-how-many-federal-agencies-exist

Dessai, S. & M. Hulme. 2004. Does Climate Adaptation Policy Need Probabilities? *Climate Policy* 4(2): 107–128.

Dunlap, R.E. 2008. The New Environmental Paradigm Scale: From Marginality to Worldwide Use. *Journal of Environmental Education* 40(1): 3–18.

Feiock, R.C. 2013. The Institutional Collective Action Framework. *Policy Studies Journal* 41(3): 397–425.

Feiock, R.C., I.W. Lee, H.J. Park, & K. Lee. 2010. Collaboration Networks among Local Elected Officials: Information, Commitment, and Risk Aversion. *Urban Affairs Review* 46(2): 241–262.

Feiock, R.C. & J.T. Scholz. 2009. Self-Organizing Governance of Institutional Collective Action Dilemmas: An Overview. In *Self-Organizing Federalism: Collective Mechanisms to Mitigate Institutional Collective Action Dilemmas*, edited by R.C. Feiock & J.T. Scholz (pgs. 3–32). Cambridge: Cambridge University Press.

Fiorino, D.J. 2000. Innovation in U.S. Environmental Policy: Is the Future Here? *American Behavioral Scientist* 44(4): 538–547.

Fowler, L. 2019. Obstacles and Motivators for Partnership Formation in a Multidimensional Environment. *Politics & Policy* 47(2): 267–299.

Grundmann, R. 2016. Climate Change as a Wicked Social Problem. *Nature Geoscience* 9: 562–563.

Gupta, J. 2010. A History of International Climate Change Policy. *Wiley Interdisciplinary Reviews: Climate Change* 1(5): 636–653.

Hardin, G. 1968. The Tragedy of the Commons. *Science* 162(3859): 1243–1248.

Herweg, N., N. Zahariadis, & R. Zohlnhofer. 2018. The Multiple Streams Framework: Foundations, Refinements, and Empirical Applications. In *Theories of the Policy Process*, 4th ed., edited by C.M. Weible & P.A. Sabatier (pgs. 17–54). Boulder, CO: Westview.

Hill, M. & P. Hupe. 2014. *Implementing Public Policy*, 3rd ed. Thousand Oaks, CA: Sage.

Hiltzik, M. 2011. *The New Deal: A Modern History*. New York: Free Press.

Hoffmann, M.J. 2011. *Climate Governance at a Crossroads: Experimenting with a Global Response after Kyoto*. Oxford: Oxford University Press.

Hsiang, S., R. Kopp, A. Jina, J. Rising, M. Delgado, S. Mohan, D.J. Rasmussen, R. Muir-Wood, P. Wilson, M. Oppenheimer, K. Larsen, & T. Houser. 2017. Estimating Economic Damage from Climate Change in the United States. *Science* 356(6345): 1362–1369.

Ingraham, P.W., P.G. Joyce, & A.K. Donahue. 2003. *Government Performance: Why Management Matters*. Baltimore, MD: Johns Hopkins University Press.

Jenkins-Smith, H.C., D. Nohrstedt, C.M. Weible, & K. Ingold. 2018. The Advocacy Coalition Framework: An Overview of the Research Program. In *Theories of the Policy Process*, 4th ed., edited by C.M. Weible & P.A. Sabatier (pgs. 135–172). Boulder, CO: Westview.

Jones, B.D. & F.R. Baumgartner. 2005. A Model of Choice for Public Policy. *Journal of Public Administration Research & Theory* 15(3): 325–351.

Kane, S. & J.F. Shogren. 2000. Linking Adaptation and Mitigation in Climate Change Policy. *Climatic Change* 45(1): 75–102.

Kennedy, D.M. 2001. *Freedom from Fear: The American People in Depression and War, 1929–1945*. New York: Oxford University Press.

King, D.A. 2004. Climate Change Science: Adapt, Mitigate, or Ignore? *Science* 303(5655): 176–177.

Kingdon, J. 1995. *Agendas, Alternatives, and Public Policies*, 2nd ed. New York: Harpers Collins.

Konisky, D.M. & N.D. Woods. 2018. Environmental Federalism and the Trump Presidency: A Preliminary Assessment. *Publius* 48(3): 345–371.

Lake, R.W. 1996. Volunteers, NIMBYs, and Environmental Justice: Dilemmas of Democratic Practice. *Antipode* 28(2): 160–174.

Lazarus, R.J. 2009. Super Wicked Problems and Climate Change: Restraining the Present to Liberate the Future. *Cornell Law Review* 5(8): 1153–1234.

Letcher, T.M. (editor). 2016. *Climate Change: Observed Impacts on Planet Earth*, 2nd ed. Waltham, MA: Elsevier.

Levin, K., B. Cashore, S. Bernstein, & G. Auld. 2012. Overcoming the Tragedy of Super Wicked Problems: Constraining Our Future Selves to Ameliorate Global Climate Change. *Policy Sciences* 45(2): 123–152.

Liebschutz, S.F. 1991. *Bargaining Under Federalism: Contemporary New York*. Albany, NY: State University of New York Press.

Lindblom, C.E. 1959. The Science of "Muddling Through". *Public Administration Review* 19(2): 79–88.

Lipsky, M. 2010. *Street-level Bureaucracy: Dilemmas of the Individual in Public Services*, 30th anniversary ed. New York: Russel Sage Foundation.

MacDonald, G.M. 2010. Water, Climate Change, and Sustainability in the Southwest. *Proceedings of the National Academy of Sciences* 107(50): 21256–21262.

March, J.G. & J.P. Olsen. 2010. *Rediscovering Institutions: The Organizational Basis of Politics.* New York: Free Press.

May, P.J. 1991. Reconsidering Policy Design: Policies and Publics. *Journal of Public Policy* 11(2): 187–206.

McGuire, M. 2006. Collaborative Public Management: Assessing What We Know and How We Know It. *Public Administration Review* 66(s1): 33–43.

Meier, K.J. & L.J. O'Toole. 2009. The Proverbs of New Public Management: Lessons from an Evidence-based Research Agenda. *American Review of Public Administration* 39(1): 4–22.

Milfont, T.L. & J. Duckitt. 2010. The Environmental Attitudes Inventory: A Valid and Reliable Measure to Assess the Structure of Environmental Attitudes. *Journal of Environmental Psychology* 30(1): 80–94.

Moe, T.M. 1984. The New Economics of Organization. *American Journal of Political Science* 28(4): 739–777.

New State Ice Co. v. Liebmann. 1932. Supreme Court of the United States. 285 U.S. 262.

Ney, S. & M. Verweij. 2015. Messy Institutions for Wicked Problems: How to Generate Clumsy Solutions. *Environment & Planning C: Government & Policy* 33(6): 1679–1696.

Ostrom, E. 2015. *Governing the Commons: The Evolution of Institutions for Collective Action.* Cambridge: Cambridge University Press.

Otto, F.E.L., D.J. Frame, A. Otto, & M.R. Allen. 2015. Embracing Uncertainty in Climate Change Policy. *Nature Climate Change* 5: 917–920.

Paerl, H.W. & V.J. Paul. 2012. Climate Change: Links to Global Expansion of Harmful Cyanobacteria. *Water Research* 46(5): 1349–1363.

Pielke, R.A., C. Landsea, M. Mayfield, J. Layer, & R. Pasch. 2005. Hurricanes and Global Warming. *Bulletin of the American Meteorological Society* 86(11): 1571–1576.

Pioneer Press. 2019. *Senator Amy Klobuchar's Remarks, As Prepared for Delivery* [online]. Available at www.twincities.com/2019/02/10/amy-klobuchars-big-speech-today-heres-a-sneak-peak/

Poortinga, W., A. Spence, L. Whitmarsh, S. Capstick, & N.F. Pidgeon. 2011. Uncertain Climate: An Investigation into Public Scepticism about Anthropogenic Climate Change. *Global Environmental Change* 21(3): 1015–1024.

Potoski, M. & N.D. Woods. 2001. Designing State Clean Air Agencies: Administrative Procedures and Bureaucratic Autonomy. *Journal of Public Administration Research & Theory* 11(2): 203–222.

Reed, D.S. 2014. *Building the Federal Schoolhouse.* Oxford: Oxford University Press.

Rittel, H.W.J. & M.W. Webber. 1973. Dilemmas in a General Theory of Planning. *Policy Sciences* 4(2): 155–169.

Ruckelshaus, W.D. 1985. Environmental Protection: A Brief History of the Environmental Movement in America and the Implications Abroad. *Environmental Law* 15(3): 455–469.

Scannell, L. & R. Gifford. 2011. Personally Relevant Climate Change: The Role of Place Attachment and Local versus Global Message Framing in Engagement. *Environment & Behavior* 45(1): 60–85.

Schneider, A.L., H. Ingram, & P. deLeon. 2014. Democratic Policy Design: Social Construction of Target Populations. In *Theories of the Policy Process*, 3rd ed., edited by P.A. Sabatier & C.M. Weible (pgs. 105–150). Boulder, CO: Westview.

Shwom, R.L., A.M. McCright, S.R. Brechin, R.E. Dunlap, S.T. Marquart-Pyatt, & L.C. Hamilton. 2015. Public Opinion on Climate Change. In *Climate Change &*

Society: Sociological Perspectives, edited by R.E. Dunlap & R.J. Brulle (pgs. 269–299). New York: Oxford University Press.

Stein, M. 2009. *How the States Got Their Shapes*. New York: Harper.

Stone, D. 2011. *Policy Paradox: The Art of Political Decision Making*, 3rd ed. New York: W.W. Norton & Co.

Svara, J.H. 1998. The Politics-Administration Dichotomy Model as Aberration. *Public Administration Review* 58(1): 51–58.

Svara, J.H. 2008. Beyond the Dichotomy: Dwight Waldo and the Intertwined Politics-Administration Relationship. *Public Administration Review* 68(1): 46–52.

Sweet, W. 2016. *Climate Diplomacy from Rio to Paris: The Effort to Contain Global Warming*. New Haven, CT: Yale University Press.

Twain, M. 1984. *Life on the Mississippi*. New York: Penguin Books.

Urwin, K. & A. Jordan. 2008. Does Public Policy Support or Undermine Climate Change Adaptation? Exploring Policy Interplay across Different Scales of Governance. *Global Environmental Change* 18(1): 180–191.

U.S. Environmental Protection Agency. 2019. *The Mississippi/Atchafalaya River Basin (MARB)* [online]. Available at www.epa.gov/ms-htf/mississippiatchafalaya-river-basin-marb

Vasi, I.B. & M. Macy. 2003. The Mobilizer's Dilemma: Crisis, Empowerment, and Collection Action. *Social Forces* 81(3): 979–998.

Waldo, D.W. 2007. *The Administrative State: A Study of the Political Theory of American Public Administration*. New Brunswick, NJ: Transaction Publishers.

Waterman, R.W. & K.J. Meier. 1998. Principal-Agent Models: An Expansion? *Journal of Public Administration Research & Theory* 8(2): 173–202.

Watts, M. 2017. Cities Spearhead Climate Action. *Nature Climate Change* 7: 537–538.

Weber, E.U. & P.C. Stern. 2011. Public Understanding of Climate Change in the United States. *American Psychologist* 66(4): 315–328.

Webster, M., C. Forest, J. Reilly, M. Babiker, D. Kicklighter, M. Mayer, R. Prinn, M. Sarofin, A. Sokolov, P. Stone, & C. Wang. 2003. Uncertainty Analysis of Climate Change and Policy Response. *Climatic Change* 61(3): 295–320.

Wilbanks, T.J. & R.W. Kates. 1999. Global Change in Local Places: How Scale Matters. *Climatic Change* 43(3): 601–628.

Wilson, J.Q. 1989. *Bureaucracy: What Government Agencies Do and Why They Do It*. New York: Basic Books.

Wilson, S.M., R. Richard, L. Joseph, & E. Williams. 2010. Climate Change, Environmental Justice, and Vulnerability: An Exploratory Spatial Analysis. *Environmental Justice* 3(1): 13–19.

Wood, B.D. & J. Bohte. 2004. Political Transaction Costs and the Politics of Administrative Design. *Journal of Politics* 66(1): 176–202.

Wood, B.D. & R.W. Waterman. 1991. The Dynamics of Political Control of the Bureaucracy. *American Political Science Review* 85(3): 801–828.

Woods, N.D. & M. Potoski, M. 2010. Environmental Federalism Revisited: Second-Order Devolution in Air Quality Regulation. *Review of Policy Research* 27(6): 721–773.

Wright, D.S. 1988. *Understanding Intergovernmental Relations*, 3rd ed. Belmont, CA: Wadsworth.

2

A BRIEF HISTORY OF ENVIRONMENTAL FEDERALISM

Shared power is a canonical feature of federalism around the world, and in the US, the balance of powers between national and state governments has evolved over two centuries. But the core assumption of this federal relationship has always remained the same: in order to satisfy the competing demands of one nation, institutions must allow local communities to determine their own best interests but also ensure that those interests do not conflict with the greater good of society as a whole. In making this assumption, the Founding Fathers embedded in the American system a sense of both local autonomy and a responsibility to national goals. However, "the best laid schemes of mice and men often go awry," and local interests are at odds with national goals time and time again. By operating at the intermediate-level, states serve as linchpins in this system by rectifying the national with the local. But this also means that states can send ripples through the federalist system in how they choose to deliver public services. For instance, if states fail to live up to their role, it leaves a power vacuum for other governmental units to fill. On the other hand, states may also decide that their role needs to be bigger, and they challenge national or local governments for control.

In order to understand environmental federalism in the 21st century, we first need to consider how we got here. Certainly, American government and society is much different today than it was in 1776. For that matter, there are fundamental differences between our government today and our government in 1976. In terms of environmental federalism, we have moved from subnational-dominated policy spheres to top-down environmental leadership to an era of contested federalism as state and national governments jockey for control of the policy agenda. This evolution has been largely shaped by the emerging policy challenges that highlight cleavages in institutions and failures to effectively manage problems. Consequently, there is an ongoing push and pull between national, state, and local governments concerning who is responsible for mitigating environmental problems, which has only been magnified as the environment has become a more important political issue. How this system evolved provides the context to understanding why environmental federalism is organized as such in the modern era and whether it can continue to evolve. As such, this chapter will examine how the evolution of federalism has coincided with and influenced the development of environmental institutions.

Foundations (1776 to 1929)

The summer of 1787 was sweltering hot in Philadelphia as 55 men packed a cramped room in the Pennsylvania State House (now Independence Hall) to debate the future of America. The Articles of Confederation had been around for a decade and pulled 13 colonies into one nation, but it had become apparent that the Articles were an untenable foundation for a national government (Vile, 2005). Foretelling the future, it was an unexpected challenge that highlighted institutional limits and forced the Founding Fathers to rethink how power was shared between sovereign governments. Flashback to Springfield, Massachusetts, nine months before the Constitutional Convention. Shay's Rebellion breaks out in response to public outcry over taxation (a common theme in American history). The national government finds itself without the power to sufficiently finance troops, and other state governments have little interest in a problem outside of their borders. The state of Massachusetts is forced to raise its own army in order to put down the rebellion (Richards, 2002). It is now clear that a loose association between the former colonies will not allow a national government to build the institutions it needs to meet the challenges ahead.

Our Founding Fathers debated many things in that steamy room in Philadelphia; the New Jersey and Virginia Plans may be the most famous (Ketcham, 2003; Vile, 2005). But one conflict that is often overlooked is how much sovereignty states would have to sacrifice in order to form a more perfect union. Let us not forget that in declaring independence from Great Britain, the colonies became sovereign governments unto themselves, so joining together meant that they would have to give up power. On one side, the Federalists advocated for a strong central government that would have a perspective on national affairs and the capacity to coordinate policy across states. On the other side, the Anti-Federalists, driven by fears that rural views and local voices would be drowned out by city concerns and national audiences, advocated for a continuation of a weak central government and strong state governments. Although the Federalists tentatively won and the Constitution formed the basis of a new union where the national government would wield supremacy over state governments, the debate was hardly settled (Ketcham, 2003; Lowi, 2006; Zimmerman, 2008). While isolated political skirmishes broke out here and there, the ensuing years saw relative peace between national and state governments, as both sides were willing to divvy up authority and stay out of each other's respective policy spheres. In turn, the Supreme Court settled disputes in order to uphold this dual federalism doctrine (Wright, 1988; Lowi, 2006; Zimmerman, 2008).

In its infancy, America was still a largely agrarian-based economy, and nearly 95 percent of the population lived in the countryside. Few economic or social activities required crossing state lines, so institutions focusing on intra-state regulation of education, public health, land use, or morality were rarely challenged. Although the Marshall Court made several landmark rulings that clarified the distinction between intra-state and inter-state commerce in cases such as McCulloch v. Maryland (1819) and Gibbons v. Ogden (1824), this doctrine would remain largely stable until the 1860s when slavery would push a union based on shared sovereignty to its limits (Wright, 1988; Peterson, 1995; Lowi, 2006; Zimmerman, 2008). Although some debate the causes of the Civil War, at its heart, the conflict centered over who controlled the fate of slaves in America. The bloodiest war in American history decided a debate that neither politicians nor lawyers could: the national government had dominion over the states (and the coercive power to enforce it), whether they liked it or not (Elazar, 1971; Spaulding, 2003; Versluis, 2007). But the true transformation of federalism happened through institutional change, not on the battlefield.

While the death knell of dual federalism would not ring until the 1930s, the post-Civil War period began its slow demise as states lost more and more ground to the national government (Lowi, 2006; Zimmerman, 2008). Emboldened by war, the national government began expanding. The National Banking Acts of 1863 and 1864 created a national banking system. The Morrill Land-Grant Act of 1862, and later of 1890, provided funding to establish universities. Later, the Interstate Commerce Act of 1887 would tighten the national government's grip on the economy (Lowi, 2006; Zimmerman, 2008). With the Pendleton Act of 1883 and the creation of the Interstate Commerce Commission in 1887, the national government would also begin to build administrative capacity (Van Riper, 1976; Rohr, 1986). Slowly, the national government crept into the policy spheres that states once dominated. This coincided with both rapid industrialization and urbanization in America, with people leaving farms for factories in cities. Social and economic activities began to connect states across America, and isolated communities were beginning to disappear. National regulatory coordination was essential to spurring this growth and removing barriers to economic opportunities as lands in the West opened up. This economic growth and urbanization also changed the way in which American society interacted with the environment, which incited a growing movement to conserve and preserve natural resources.

The origins of the American environmental movement find themselves in the sportsmen and naturalists who saw the forces of rapid industrialization and urbanization degrading their playgrounds and field laboratories in the late

1870s (Kline, 2011). But it was not until the progressive movement of the 1890s and 1900s emerged that conservationists began to have a distinct political voice, and a captive audience in President Theodore Roosevelt's White House. Prior to the 20th century, environmental policy was largely left to the states, which developed piecemeal regulations only when dire situations emerged. Local governments were left to settle disputes over natural resources, but these were largely treated in the same way as disputes over other goods. As such, subnational policies were largely devoid of long-term planning or comprehensive regulatory institutions. To this end, Roosevelt would break new ground during his time in the White House by creating some of the first environmental institutions in the US with the signing of several major pieces of legislation to create and protect natural resources, such as the Antiquities Act of 1906, as well as the creation of the first National Parks, National Monuments, and National Forests. However, other than the US Forest Service, there were still few public agencies dedicated to the protection of the environment, and states still dominated environmental policy (Brinkley, 2009; Kline, 2011).

New Deal, New Era (1929 to 1980)

It would be Roosevelt's New Deal that would begin a new era in national-state relationships (Wright, 1988; Lowi, 2006; Zimmerman, 2008). As the Great Depression undermined more than a century of American economic progress, Franklin Delano Roosevelt was elected to the White House on a platform that promised sweeping changes to the way government participated in society (Kennedy, 2001; Hiltzik, 2011). While his 1933 inauguration speech is best remembered as "the only thing we have to fear is fear itself" speech, he would also foreshadow the change in federalism that was needed in order to implement the New Deal to which he had pledged himself and this nation. Roosevelt called for federal, state, and local governments to work together in order to unify

> relief activities which today are often scattered, uneconomical, and unequal . . . [which] can be helped by national planning for and supervision of all forms of transportation and of communication and other utilities which have a definitely public character.
>
> *(Avalon Project, 2019)*

Although the Supreme Court initially knocked down New Deal proposals, they changed direction after the Judicial Procedures Reform Bill of 1937 threatened to expand the court and allow Roosevelt to pack it with judges friendly to his cause. This in turn kicked off a new era of federalism that transformed our institutions from separate and independent to nationally coordinated and interdependent (Lowi, 2006; Zimmerman, 2008).

While modern scholars refer to this as "cooperative federalism," that name can be misleading. National, state, and local governments were treated as a single system under the coordination of federal politicians and bureaucrats (Agranoff & McGuire, 2001). States were expected to cooperate, and, if they did not, they were bypassed or coerced. Local governments were treated with new respect and helped connect national policies to local communities. The formally distinct lines separating layers of government disappeared, and their authorities began to overlap. Federal grants to state and local governments created resource dependencies. Congress used the Commerce clause to create new programs that further stretched its footprint. The national government dominated through incentives, mandates, or coercion, whichever was most effective. Despite numerous legal challenges, the Supreme Court ruled in favor of the national government on every federalism case between 1937 and 1995 (Wright, 1988; Lowi, 2006; Zimmerman, 2008). The war effort from 1939 to 1945 and the post-war economic boom only further empowered the national government with resources, capacities, and international prestige. But this boom also ushered in new problems and a new era in environmental governance as the national government finally began to take environmental issues seriously.

Although the Refuse Act of 1899 was the first national law protecting the environment, the Federal Water Pollution Control Act (FWPCA) of 1948 was the first major federal initiative that incorporated national, state, and local governments into a system for environmental protection. Up to that point in US history, waterways were largely treated as waste disposal sites or open sewage systems, which led to rivers and streams around major cities that were incapable of supporting life. In response, Congress adopted a comprehensive set of water quality programs to protect interstate waterways from being used as dumping grounds and provided enforcement funding to state and local governments (Fowler, 2014; Copeland, 2016). The Air Pollution Control Act (APCA) of 1955 may be a more important milestone for environmental federalism though, as it declared air pollution to be a public health issue in which the national government had a vested interest but left primary responsibility for controlling air quality to the states. The APCA also provided federal funding for research on air pollutant emissions and directed that findings be distributed to states to inform their activities (Fowler, 2014). Whether this initial legislation was successful is questionable, but the FWPCA and the APCA form the historical and institutional basis for modern water and air quality programs.

In the ensuing years, several factors grabbed the public's attention and focused it on the environment, further spurring the environmental movement and new legislation. Socially, American citizens were becoming more critical of existing power structures and more civically engaged. At the same time, increased outdoor recreation was bringing people into closer contact with environmental degradation, and economic growth in the post-World War II era led to a restructuring of social values around the quality of life. Existing environmental groups began to reorganize to attract support from a broad-based coalition of interests, and new advocacy groups emerged to focus on specific environmental issues within their communities (Dunlap & Mertig, 2013). People were also becoming more informed about the impacts of pollutants on human health and wildlife as scientific knowledge developed. For instance, Rachel Carson's *Silent Spring* (1962) would be a watershed moment in the environmental movement as it drew national attention to negative impacts of pesticides on the environment and the role of industry in spreading misleading information.

Additionally, several major man-made disasters highlighted the consequences of weak regulations. For instance, in November 1968, a coal mine explosion near Farmington, West Virginia, left 78 miners dead and raised serious questions about mine safety (Imbrogno, 2018). Only two months later, in January 1969, a blowout on an offshore oil rig near Santa Barbara caused between 80,000 and 100,000 barrels of crude oil to spill into the Santa Barbara Channel; this would be the largest oil spill in US waters until the Exxon *Valdez* disaster in 1989 (Mai-Duc, 2015). By the time a spark from a passing railcar ignited an oil slick on the Cuyahoga River in June 1969, national media was attuned to the significance of these incidents. Although the Cuyahoga River has caught fire at least a dozen times in history, a *Time* magazine article on pollution in America's waterways fixed national attention on it, despite the relatively minor damages (about $50,000) (Boissoneault, 2019). Collectively, these events would culminate in the first Earth Day held on April 22, 1970, in which thousands of demonstrators publicly expressed their support for environmental protection (Dunlap & Mertig, 2013). Policymakers could no longer ignore that Americans were becoming more concerned and more aware of environmental dangers and were demanding more aggressive regulations (Downs, 1996).

In response, there was a virtual explosion of national legislation in the 1960s, such as the CAA of 1963, the Air Quality Act of 1967, the Water Quality Act of 1965, and the Solid Waste Disposal Act of 1965. Unfortunately, these laws were relatively toothless and did little to spur serious environmental action on the part of the federal government. Subsequently, the real culmination of early efforts was the National Environmental Protection Act (NEPA) of 1969, which finally established a national policy to

> encourage productive and enjoyable harmony between man and his environment; to promote efforts which will prevent or eliminate damage to the environment and biosphere and stimulate the health and welfare of man; to enrich the understanding of the ecological systems and natural resources.
>
> *(Council on Environmental Quality, 2020)*

In principal, it declared environmental quality as a public value of the American people and required national agencies to consider environmental impacts in their decision-making. It also created the Council of Environmental Quality, which would be the first formal national advisory group dedicated to counseling the president on matters of environmental policy. President Richard Nixon also took the extraordinary step of creating the EPA via executive order to reorganize the environmental responsibilities of the national government under one agency (Train, 1996; Flippen, 2000). For the first time in history, there was a federal agency dedicated to environmental protection.

Under Nixon and his successor, Gerald Ford, the EPA's mission would quickly expand and major environmental laws would mark the beginning of a modern "green" state. Nixon signed into law the CAA of 1970, the CWA of 1972, the Federal Insecticide, Fungicide, and Rodenticide Act (FIFRA) of 1972, and the Endangered Species Act of 1973. Ford also signed into law the Safe Drinking Water Act (SDWA) of 1974 and the RCRA of 1976; later, Jimmy Carter added the Comprehensive Environmental Response, Compensation, and Liability Act of 1980 (i.e., Superfund) (Kraft, 2000). Consequently, between the adoption of the CAA of 1963 and the founding of the EPA, national spending and grants to states for pollution control and abatement, and natural resources, rose. Spending during this period would be dwarfed after the EPA was founded and programs expanded (see Table 2.1) (Office of Management & Budget (OMB), 2019). In response, state spending on natural resources followed a similar pattern and also reached historic levels by 1980 (Census, 2019). Furthermore, several states created or reorganized environmental missions under new agencies,

TABLE 2.1 Federal and State Environmental Spending, 1963 to 1980

Category	1963	1970	1980	% Change From CAA of 1963 to EPA Founding	% Change From EPA Founding to Reagan
Federal Spending					
Natural Resources	$16,041	$17,225	$36,673	7.39%	112.90%
EPA	–	$2,158	$14,827	–	587.06%
Pollution Control and Abatement	$620	$2,158	$14,581	248.10%	575.65%
Total	$793,239	$1,099,547	$1,563,813	38.61%	42.22%
Federal Grants-in-Aid to State and Local					
Natural Resources	$1,162	$2,310	$14,192	98.86%	514.43%
EPA	–	$1,090	$12,181	–	1017.23%
Total	$61,298	$135,245	$241,883	120.64%	78.81%
State Spending					
Natural Resources	$841	$12,493	$14,793	1385.75%	18.41%
Total	$282,062	$478,009	$972,096	69.47%	103.36%

Note: The Consumer Price Index (CPI) is used to convert nominal dollars as reported by OMB and the US Census into real 2010 dollars for the purpose of comparison over time.
* Dollar amounts shown in the table are in millions

such as the New Jersey Department of Environmental Protection, Ohio Environmental Protection Agency, Oregon Department of Environmental Quality, or Pennsylvania Department of Environmental Protection. In other words, national and state governments were finally taking environmental policy seriously and creating the necessary institutions to manage it.

Although a bipartisan coalition in Congress had come to agree on basic regulatory structures needed for environmental protection through this legislation, a nagging question still remained on how to implement it (Kraft, 2000). As early legislation had been based on the principle that states retained the rights to regulate the environment and national agencies ultimately lacked the capacity to make these policies work at the local-level, it was decided that states would be a key conduit for translating these new programs into practice. Consequently, most major national environmental programs (e.g., CAA, CWA) are based on a partial preemption system that allows the national government to delegate primary implementation and enforcement responsibility (i.e., primacy) to state governments but retain the authority to determine the adequacy of state actions. Conventionally, this plays out as the EPA setting broad guidelines and standards that govern state programs, and states developing rules, regulations, and administrative procedures to match national goals to local technical, economic, and sociopolitical challenges. States are allotted discretion in designing programs and may exceed national standards if they deem it appropriate. Ultimately, the EPA retains the authority to preempt all or parts of state programs that it deems inadequate and to directly implement programs in the absence of adequate state programs (Woods, 2006; Woods & Potoski, 2010).

According to the EPA's founding administrator, William Rucklehaus, this was all based on the belief

> that the states had enough interest and infrastructure to enforce these laws. They also had this "gorilla in the closet" – that is, the federal government, which could assume control if the state authorities proved too weak or inept to curb local polluters . . . so long as [the gorilla] didn't come out of the closet and we let the states alone, the gorilla could help induce compliance.
>
> *(EPA, 1993)*

Moreover, James Seif, former Region 3 administrator, contends:

> The reality was that nobody thought that EPA could, if we had yanked the program as the saying went, do it ourselves. It was a political impossibility in West Virginia, a bureaucratic impossibility from Philadelphia to Charleston with no local knowledge, and no people, and so on.
>
> *(EPA Alumni Association, 2015)*

In other words, the EPA believed that states were key partners in environmental regulation, but they had to create mechanisms to ensure cooperation or risk the entire system collapsing. For starters, the process of delegation did not automatically happen, and states had to request primacy and prove they had the capacity to implement programs. Of course, some states "were way ahead of EPA in terms of capacity, zeal, and interest" (EPA Alumni Association, 2015), but others required both carrots and sticks to encourage compliance.

Consequently, while states had discretion in theory during this period, in reality, they were severely constrained by national regulations, mandates, and inducements. There was also the lingering threat that the EPA would swoop in and take over programs if states did not live up to expectations. From the EPA's perspective, they were reluctant to place so much authority in state hands without tools to ensure they could direct agencies to achieve environmental goals. Of course, "this oversight created a very, very difficult period between the EPA and the states. The states thought [the EPA] dictated too much, were too intrusive" (EPA, 1993). In turn, states used control over these programs to negotiate funding and program adjustments while wielding their own threats of dropping programs and letting the EPA figure out what to do in the aftermath (Derthick, 1987; Crotty, 1987). Despite these conflicts, states were still willing to take on primacy. Of the 13 sections of the CAA, CWA, FIFRA, and SDWA for which responsibility could be delegated in 1980, 21 states had primacy for a majority and every state had primacy for at least three (Crotty, 1987). But not everyone saw national dominion over states as institutional progress, especially as the national government created more environmental programs that were unpopular among industry leaders and political conservatives.

New Federalism, New Power Dynamics (1980 to 2000)

Like the summer of 1787 in Philadelphia, Pennsylvania, the summer of 1980 was sweltering hot in Philadelphia, Mississippi, as Ronald Reagan took the stage at the Neshoba County Fair to kick off his general election campaign for the presidency. Reagan's rise to national political prominence was on the back of a new brand of conservatism, but institutional legacies and politics of the past were not lost on him:

> I believe that we've distorted the balance of our government today by giving powers that were never intended in the Constitution to that federal establishment. And if [elected President], I'm going to devote myself to trying to reorder those priorities and to restore to the states and local communities those functions which properly belong there.
>
> *(Neshoba County Democrat, 2007)*

While there may have been subtly in his dog-whistle politics in regards to "state's rights," Reagan's intentions to reorder American federal institutions and give power back to the states was obvious, and it was an intention that he stuck with during his time in the White House. Although it would be another decade and a half before the Supreme Court made rulings, such as United States v. Lopez (1995) and United States v. Morrison (2000), that reined in federal regulatory power, Reagan's Devolution Revolution would be a turning point (Nathan & Doolittle, 1987; Lowi, 2006; Zimmerman, 2008).

The new approach to federalism was for the national government to establish standards or guidelines and for states to implement policies. In turn, states would be released from rigid top-down controls and allowed to develop policies and implementation strategies that fit their unique jurisdictional needs. In effect, the national government was taking a step back from many of the social and environmental policies pioneered over the previous decades and forcing the states to fill that policy space, whether they liked it or not. Although New Federalism finds its roots in the block grants given out during the Nixon administration, over the course of his presidency, Reagan devolved authorities over programs in almost every major domestic policy area to states so they could create programs that fit their unique social, economic, political, and technical circumstances. But, more importantly, it was part of a larger push to reduce the size and scope of the national government (at least domestically) and replace it with a subnational government. While Reagan's administration spent lavishly on defense, it also worked to dismantle national programs and the bureaucracies that managed them in accordance with supply-side economic theories (Nathan & Doolittle, 1987; Lowi, 2006; Zimmerman, 2008).

Specific to environmental policy, Lester (1986) identifies three guiding principles for Reagan's New Federalism: "1. regulatory reform, involving extensive use of cost-benefit analysis in determining the value of environmental regulations and programs. 2. reliance as much as possible on the free market to allocate resources; and, 3. decentralization or environmental federalism, shifting responsibilities for environmental protection to state and local governments whenever feasible" (p. 151). To this end, states were allotted a greater degree of programmatic flexibility while being granted more deference in the development of national policies. In practice, this resulted in grants to states being defunded, legislation that preempted states and confined them to certain parameters, and the amplification of unfunded mandates to ensure states did not fully retreat from environmental protection (Lester, 1986; Zimmerman, 1991). Possibly more than any other policy area than welfare, Reagan's administration was particularly antagonistic towards the environment, and many argue that his environmental record is the worst of any modern president. Criticisms of Reagan's approach to allowing national agencies to wither and pushing environmental responsibilities to the states came from both sides; environmentalists argued that it was a lost eight years for environmental policy, while conservatives argued that it was a missed opportunity to reform the national bureaucracy into a system that could protect the environment with an affordable price tag (Shabecoff, 1989).

In February 1990, George H.W. Bush reaffirmed Executive Order 12612, which required national agencies to conduct federalism assessments of proposed policies to determine how to balance national and state authority. This, in effect, signaled that Bush would continue to rely on New Federalism as a governing philosophy for his policy agenda. However, he would also take a more pro-environmental approach than his predecessor. During his campaign, Bush was a self-declared "environmentalist." As such, he was willing to break from the precedent set by Reagan and allow the "EPA to take a fairly aggressive stance against state actions affecting the environment" (Bowman & Pagano, 1990, p. 9). Likely, the biggest milestone of his tenure was signing into law the CAA amendments of 1990, which modernized air quality regulations and expanded the authorities of both the EPA and the states. The CAA of 1990 also created a market-based system for encouraging reduced emissions related to acid rain, which would be one of the first policy experiments in cap-and-trade or emissions trading systems. Although Bush was not a proponent of early action on global warming, his administration did participate in the Rio Earth Summit, the world's first conference on climate change (Bowman & Pagano, 1990). Despite criticisms, Bush's term is generally regarded as a step forward after a decade of stagnation in environmental policy.

Interestingly, Ronald Reagan would be the first president in modern US history who would oversee a decrease in environmental spending. With a few exceptions, the national government would reduce spending and grants to states for every major environmental program (see Table 2.2) (OMB, 2019). Although national spending and grants to states were increasing in some areas, the national government was retreating from environmental policy, as well as some social policy areas. In response, states were forced to increase their own environmental spending in order to make up for federal retrenchment (Census, 2019). On the other hand, the Bush administration would increase environmental spending, albeit not to pre-Reagan levels, during his one and only term (OMB, 2019). In general, trends in environmental spending reflect both the importance placed on environmental policy during this era and the back and forth between national and state governments. That is, when national spending was reduced, states had to compensate, and when national spending returned, states slowed their own spending. Notably, national spending also shrank while state spending grew under the Clinton administration, but this was likely due to a new approach to environmental federalism.

Although Bill Clinton's administration would adhere to the broader principles of New Federalism, his larger movement to reinvent government would cast the relationship between national and state governments in a different light. In general, this approach relied on the federal government "leading" by establishing national goals and rigorous performance standards, reducing unfunded mandates on subnational governments, coordinating collaborative problem-solving efforts, and increasing national fiscal responsibility. On the

TABLE 2.2 Federal and State Environmental Spending, 1988 to 2000

Category	1988	1992	2000	% Change Under Reagan	% Change Under Bush	% Change Under Clinton
Federal Spending						
Natural Resources	$26,913	$31,086	$31,661	−26.61%	15.50%	1.85%
EPA	$8,978	$9,248	$9,146	−39.45%	3.00%	−1.09%
Pollution Control and Abatement	$8,907	$9,439	$9,364	−38.92%	5.98%	−0.79%
Total	$1,961,981	$2,147,190	$2,265,338	25.46%	9.44%	5.50%
Federal Grants-in-Aid to State and Local						
Natural Resources	$6,907	$6,107	$5,816	−51.33%	−11.59%	−4.73%
EPA	$5,336	$4,722	$4,419	−56.19%	−11.52%	−6.40%
Total	$212,604	$276,751	$362,001	−12.09%	30.17%	30.80%
State Spending						
Natural Resources	$15,317	$14,022	$20,219	3.55%	−8.46%	44.19%
Total	$796,612	$1,087,877	$1,372,787	−18.05%	36.56%	26.19%

Note: The CPI is used to convert nominal dollars as reported by OMB and the US Census into real 2010 dollars for the purpose of comparison over time.

other hand, states would be granted greater flexibility to design programs that fit their communities, so they could play their classic role as laboratories of democracy (Galston & Tibbetts, 1994). While those principles sound similar to previous administrations, the key difference here is that Clinton did not plan for the national government to retreat from policy, but rather to find ways to collaborate with subnational governments and make policy work at the local-level. In other words, zero-sum game playing between national and state governments would be replaced with cooperation and overlapping efforts. In theory, this represents a system that combines both top-down and bottom-up elements by creating a framework for state actions to achieve defined national goals; in practice, it required national-state cooperation to be successful (Scheberle, 2005).

Clinton's Reinventing Government movement also shifted the broader regulatory philosophy of the federal government. In general, it advocated for federal agencies to reconsider how existing regulations created negative externalities and institutional barriers to problem solving, as well as alternatives to direct regulation, such as market-based mechanisms to incentivize behavior or reduce information asymmetries. To this end, the EPA developed the Common Sense Initiative and Project XL to work with stakeholders to streamline administrative and regulatory processes in order to lower costs and make compliance easier on industries. Additionally, agencies were also required to consult with state and local governments in order to avoid "one-size-fits-all" regulations that disadvantaged specific jurisdictions (Galston & Tibbetts, 1994). In response, the EPA began pursuing Performance Partnership Agreements that allowed the EPA and states to negotiate the specific terms of their partnerships, including programmatic requirements, the EPA's oversight activities, and program evaluation criteria, and opened up new grant opportunities that sought to address environmental priorities of specific states (Kraft & Scheberle, 1998). Notably, in 1993, the Environmental Council of the States (ECOS) was founded by environmental officials from all 50 states in order to protect state interests as national environmental regulations were reorganized (Rabe, 2007). In sum, by the end of the Clinton administration, the states were playing much more active roles in the federal system than they were in 1980.

Bush and Beyond (2000 to 2019)

George W. Bush's presidency started with expectations that trends in decentralized and cooperative approaches to environmental policy would continue. While Governor of Texas, Bush had endorsed major legislative forms for renewable energy and air quality, and his campaign rhetoric emphasized providing state and local governments with control over environmental decisions affecting their communities. Early environmental policy initiatives, such as the Small Business Liability and Revitalization Act of 2002, seemed to conform to these expectations. But cuts in national environmental spending undermined state and local engagement during policy implementation, and early support for Bush's environmental agenda began to dissipate. In response, Bush, and a cohort of his key advisors led by Vice President Dick Cheney, moved to centralize decision-making and marginalize threats to their agenda. In many cases, this relied on reinterpretations of existing legislation, financial incentives to encourage compliance from states, and expanded mandates to decrease their capacities to pursue their own priorities (Rabe, 2007). Although Bush and his successor fundamentally differed in their approach to the environment, President Barack Obama would largely follow Bush's lead in pursuing his own environmental agenda (Konisky & Woods, 2016).

For Obama, the environment, particularly as it relates to climate change, was a major issue both as candidate and president. Obama's environmental legacy is most readily defined by either conflict with Congressional Republicans and industry who accused him of going too far, or criticisms from members of environmental advocacy coalitions who accused him not going far enough. But his key policy initiatives relied on states for implementation, which sparked conflicts. For instance, while the American Recovery and Reinvestment Act of 2009 was chiefly focused on stimulating economic growth, it was also designed to achieve a variety of policy goals, including those related to energy and climate change. Although pass-through grant funds ended up in the hands of local governments, nonprofits, or private contractors, states were given discretion on distribution, which resulted in some dragging their feet on implementation. On the other hand, while the Clean Power Plan gave states discretion to come up with energy plans for emission reductions, if states failed to do so, the EPA was authorized to replace state plans with their own, which they frequently did (Konisky & Woods, 2016). In both cases, this drew the ire of state leaders, who felt the Obama administration was overreaching its authority. Early evidence suggests Donald Trump is also a fan of achieving his agenda through executive orders, such as withdrawing the US from the Paris Agreement. Notably, Trump has also taken unprecedented steps to roll back environmental regulations and encourage climate change denial (Konisky & Woods, 2018).

Trends over the last three presidencies largely indicate that states are responding to attempts to recentralize environmental policy at the national-level in one of three ways. On one hand, some are taking the opportunity to retreat from environmental policy, so they can liberate themselves from a policy area that creates highly concentrated costs on select constituents. On the other hand, some are further expanding their role through innovation, and, in doing so, are attempting to take firm control over the environmental agenda. On yet another hand, others are simply complying with federal initiatives in response to the relative incentives and in order to avoid conflict. In effect, states responses have been extremely important in determining how successful presidents have been in accomplishing their environmental agendas. When states comply, presidents are largely able to take control and reshape regulatory regimes, but when states obstruct these initiatives by either disengaging or opposing, national agencies struggle to implement policies (Rabe, 2007; Konisky & Woods, 2016, 2018). Differential responses seem to be a result of how state policy actors perceive the costs and benefits of environmental policy, which is largely a continuation of trends that began in the 1970s when national programs first began to rely on states for implementation (Lester, 1995).

Nevertheless, in more recent years, states are becoming more innovative with policy in response to a lack of national leadership, while the national government seems willing to cooperate in some areas but not others. As such, the mechanisms of contemporary environmental federalism are far less defined than those of previous eras, and differences between states are leading to battle lines both vertically and horizontally. For instance, in 2003, a group of attorneys general from Northeastern states sued the Bush administration for failing to regulate greenhouse gases produced from coal-fired power plants. The eight Northeastern states were later joined by another 12 states, as well as cities and environmental groups, while a group of nine other states from the West and South publicly supported the EPA's position (Scheberle, 2005). Furthermore, while the Obama administration set new precedents for overruling and replacing state implementation plans with its own more aggressive plans, the EPA was still sued by states for not implementing the CAA aggressively enough to protect human health (Konisky & Woods, 2016). Ultimately, this highlights a clear division emerging between states, which is only amplified by attempts at horizontal policy coordination that creates regional governance without national government (Bowman, 2004; Conlan, 2006). Unfortunately, in an era of partisan polarization, national-state relationships are largely burdened by the party affiliation of elected officials (Rose & Bowling, 2015; Conlan & Posner, 2016).

Consequently, scholars have struggled to make sense of it, and find evidence of national domination in some areas and state leadership in others (Conlan, 2006; McGuire, 2006; Thompson, 2013; Konisky & Woods, 2016, 2018). Certainly, Bush and Obama's push for centralization based on different environment agendas that pulled states in opposite directions has created an ebb-and-flow to the institutions of federalism. From a broader perspective, some argue that federalism is becoming more opportunistic and strategically driven, so political rewards dictate how governments interact. Others argue that federalism is still cooperative at the local-level, where policymakers and administrators focus on problem-solving in order to serve their communities (Conlan, 2006; McGuire, 2006). The emergence of network governance and collaborative approaches to shared policy goals has further led to a more decentralized power structure and more active local governments and non-governmental organizations (NGOs) in environmental governance (Kettl, 2015). But those relationships are still very much shaped by how power is shared between national and state governments, as that power is manifested in their action or inaction to solve problems big and small. As such, even in an era of "messy" solutions, our federalist institutions tie policy governance back to the Constitution and the sovereign power of the people, making it an important point of inquiry.

References

Agranoff, R. & M. McGuire. 2001. American Federalism and the Search for Models of Management. *Public Administration Review* 61(6): 671–681.

Avalon Project. 2019. *First Inaugural Address of Franklin D. Roosevelt* [online]. Available at http://avalon.law.yale.edu/20th_century/froos1.asp

Boissoneault, L. 2019. The Cuyahoga River Fire at Least a Dozen Times, But No One Cared until 1969. *Smithsonian Magazine* [online]. Available at www.smithsonianmag.com/history/cuyahoga-river-caught-fire-least-dozen-times-no-one-cared-until-1969–180972444/

Bowman, A. O'M. 2004. Horizontal Federalism: Exploring Interstate Interactions. *Journal of Public Administration Research & Theory* 14(4): 535–546.

Bowman, A.O'M. & M.A. Pagano. 1990. The State of American Federalism 1989–1990. *Publius* 20(3): 1–25.

Brinkley, D. 2009. *The Wilderness Warrior: Theodore Roosevelt and the Crusade for America*. New York: Harper Collins.

Carson, R. 1962. *Silent Spring.* Boston, MA: Houghton Mifflin.

Conlan, T. 2006. From Cooperation to Opportunistic Federalism: Reflections on the Half-Century Anniversary of the Commission on Intergovernmental Relations. *Public Administration Review* 66(5): 663–676.

Conlan, T. & P.L. Posner. 2016. American Federalism in an Era of Partisan Polarization: The Intergovernmental Paradox of Obama's "New Nationalism". *Publius* 46(3): 281–307.

Copeland, C. 2016. *Clean Water Act: A Summary of the Law.* Congressional Research Service Report 7–5700.

Council on Environmental Quality. 2020. *National Environmental Policy Act of 1969* [online]. Available at https://ceq.doe.gov/laws-regulations/laws.html

Crotty, P.M. 1987. The New Federalism Game: Primacy Implementation of Environmental Policy. *Publius* 17(2): 53–67.

Derthick, M. 1987. American Federalism: Madison's Middle Ground in the 1980s. *Public Administration Review* 47(1): 66–74.

Downs, A. 1996. Up and Down with Ecology: The "Issue-Attention Cycle". In *The Politics of American Economic Policy Making*, 2nd ed., edited by P. Peretz (pgs. 48–59). Armonk, NY: M.E. Sharpe.

Dunlap, R.E. & A.G. Mertig. 2013. The Evolution of the U.S. Environmental Movement from 1970 to 1990: An Overview. In *American Environmentalism: The U.S. Environmental Movement, 1970–1990*, edited by R.E. Dunlap & A.G. Mertig (pgs. 1–10). New York: Routledge.

Elazar, D.J. 1971. Civil War and the Preservation of American Federalism. *Publius* 1(1): 39–58.

EPA Alumni Association. 2015. *Forty Years of Service: An Interview with James (Jim) Seif* [online]. Available at www.epaalumni.org/userdata/pdf/3CA74C4EFBBC2A5B.pdf#page=1

Flippen, J.B. 2000. *Nixon and the Environment.* Albuquerque, NM: University of New Mexico Press.

Fowler, L. 2014. Assessing the Framework of Policy Outcomes: The Case of the U.S. Clean Air Act and Clean Water Act. *Journal of Environmental Assessment Policy & Management* 16(4): 1450034.

Galston, W.A. & G.L. Tibbetts. 1994. Reinventing Federalism: The Clinton/Gore Program for a New Partnership among the Federal, State, Local, and Tribal Governments. *Publius* 24(3): 23–47.

Gibbons v. Ogden. 1824. Supreme Court of the United States. 22 U.S. 1.

Hiltzik, M. 2011. *The New Deal: A Modern History.* New York: Free Press.

Imbrogno, D. 2018. Farmington No. 9: The West Virginia Disaster that Changed Coal Mining Forever. *West Virginia Public Broadcasting* [online]. Available at www.wvpublic.org/post/farmington-no-9-west-virginia-disaster-changed-coal-mining-forever#stream/0

Kennedy, D.M. 2001. *Freedom from Fear: The American People in Depression and War, 1929–1945.* New York: Oxford University Press.

Ketcham, R. (editor). 2003. *The Anti-Federalist Papers and the Constitutional Debates.* New York: Penguin.

Kettl, D. 2015. *The Transformation of Governance: Public Administration for the Twenty-First Century*, updated ed. Baltimore, MD: Johns Hopkins University Press.

Kline, B. 2011. *First Along the River: A Brief History of the U.S. Environmental Movement*, 4th ed. New York: Rowman & Littlefield.

Konisky, D.M. & N.D. Woods. 2016. Environmental Policy, Federalism, and the Obama Presidency. *Publius* 46(3): 366–391.

Konisky, D.M. & N.D. Woods. 2018. Environmental Federalism and the Trump Presidency: A Preliminary Assessment. *Publius* 48(3): 345–371.

Kraft, M.E. 2000. U.S. Environmental Policy and Politics: From the 1960s to the 1990s. *Journal of Policy History* 12(1): 17–42.

Kraft, M.E. & D. Scheberle. 1998. Environmental Federalism at Decade's End: New Approaches and Strategies. *Publius* 28(1): 131–146.

Lester, J.P. 1986. New Federalism and Environmental Policy. *Publius* 16(1): 149–165.

Lester, J.P. 1995. Federalism and State Environmental Policy. In *Environmental Politics and Policy: Theories and Evidence*, 2nd ed., edited by J.P. Lester (pgs. 39–60). Durham, NC: Duke University Press.

Lowi, T.J. 2006. *The End of the Republican Era*. Norman, OK: University of Oklahoma Press.

Mai-Duc, C. 2015. The 1969 Santa Barbara Oil Spill that Changed Oil and Gas Exploration Forever. *Los Angeles Times* [online]. Available at www.latimes.com/local/lanow/la-me-ln-santa-barbara-oil-spill-1969-20150520-htmlstory.html

McCulloch v. Maryland. 1819. Supreme Court of the United States. 17 U.S. 316.

McGuire, M. 2006. Intergovernmental Management: A View from the Bottom. *Public Administration Review* 66(5): 677–679.

Nathan, R.P. & F.C. Doolittle. 1987. *Reagan and the States*. Princeton, NJ: Princeton University Press.

National Environmental Policy Act of 1970, Pub. L. 91–190, 83 Stat. 852. Codified as amended at 42 U.S. Code § 4321.

Neshoba County Democrat. 2007. *Transcript of Ronald Reagan's 1980 Neshoba County Fair Speech* [online]. Available at www.neshobademocrat.com/Content/NEWS/News/Article/Recording-of-Reagan-s-Fair-speech-found/2/297/13920

Office of Management & Budget. 2019. *President's Budget: Historical Tables* [online]. Available at www.whitehouse.gov/omb/historical-tables/

Peterson, P.E. 1995. *The Price of Federalism*. Washington, DC: Brookings Institutions Press.

Rabe, B. 2007. Environmental Policy and the Bush Era: The Collision between the Administrative Presidency and State Experimentation. *Publius* 37(3): 413–431.

Richards, L.L. 2002. *Shay's Rebellion: The American Revolution's Final Battle*. Philadelphia, PA: University of Pennsylvania Press.

Rohr, J.A. 1986. *To Run a Constitution: The Legitimacy of the Administrative State*. Lawrence, KS: University Press of Kansas.

Rose, S. & C.J. Bowling. 2015. The State of American Federalism 2014–2015: Pathways to Policy in an Era of Party Polarization. *Publius* 45(3): 351–379.

Scheberle, D. 2005. The Evolving Matrix of Environmental Federalism and Intergovernmental Relationships. *Publius* 35(1): 69–86.

Shabecoff, P. 1989. Reagan and Environment: To Many, a Stalemate. *Rolling Stone*, January 2.

Spaulding, N.W. 2003. Constitution as Countermonument: Federalism, Reconstruction, and the Problem of Collective Memory. *Columbia Law Review* 103(8): 1992–2051.

Thompson, F.J. 2013. The Rise of Executive Federalism: Implications for the Picket Fence and IGM. *American Review of Public Administration* 43(1): 3–25.

Train, R.E. 1996. The Environmental Record of the Nixon Administration. *Presidential Studies Quarterly* 26(1): 185–196.

United States v. Lopez. 1995. Supreme Court of the United States. 514 U.S. 549.

United States v. Morrison. 2000. Supreme Court of the United States. 529 U.S. 598.

U.S. Census Bureau. 2019. *Statistical Abstracts of the States* [online]. Available at www.census.gov/library/publications/time-series/statistical_abstracts.html

U.S. Environmental Protection Agency. 1993. *William D. Ruckelshaus: Oral History Interview* [online]. Available at https://archive.epa.gov/epa/aboutepa/william-d-ruckelshaus-oral-history-interview.html

Van Riper, P.P. 1976. *History of the United States Civil Service.* Westport, CT: Greenwood Press.

Versluis, A. 2007. Secession and American Federalism. *Modern Age* 49(3): 308–315.

Vile, J.R. 2005. *The Constitutional Convention of 1787: A Comprehensive Encyclopedia of America's Founding.* Santa Barbara, CA: ABC-CLIO, Inc.

Woods, N.D. 2006. Primacy Implementation of Environmental Policy in the U.S. States. *Publius* 36(2): 259–276.

Woods, N.D. & M. Potoski. 2010. Environmental Federalism Revisited: Second-Order Devolution in Air Quality Regulation. *Review of Policy Research* 27(6): 721–739.

Wright, D.S. 1988. *Understanding Intergovernmental Relations.* Belmont, CA: Duxbury Press.

Zimmerman, J.F. 1991. Federal Preemption under Reagan's New Federalism. *Publius* 21(1): 7–28.

Zimmerman, J.F. 2008. *Contemporary American Federalism: The Growth of National Power,* 2nd ed. Albany, NY: State University of New York Press.

3

THE POLITICS OF
ENVIRONMENTAL PROTECTION

One of the key dimensions in dictating how states participate in the federal environmental policy system is the internal political will to commit to environmental protection (Lester, 1995). In some states, the political context is particularly amenable to rigorous environmental regulations, while in other states, policymakers will hardly consider such initiatives. So, how do we know to what degree political incentives for environmental protection exist within individual states? Previous scholars have considered this question and how it helps explain patterns of policymaking and environmental conditions (e.g., Mazur & Welch, 1999). While there are numerous potential ways to answer that question, in this chapter, we look at three facets: public opinion, advocacy groups, and intergovernmental competition. Specifically, these facets provide insights into public beliefs, the influence of advocacy groups, and how states compare to their competitors. Collectively, they are indicative of the context in which environmental policy choices are made and provide a basis to make inter-state comparisons. The goal in examining these sources of political incentives is to determine in which states we should expect to find high degrees of commitment to environmental protection, and in which states low degrees.

Political Incentives

Political incentives are the rewards available to decision-makers, and the more likely decision-makers are to be rewarded for a certain behavior, such as making a commitment of state resources to environmental protection, the more motivated they will be to do so. Now, let us consider specifically the political incentives for environmental protection. Politics is about who gets what, when and how, so policy actors respond to what they get, when and how

(Lasswell, 1936). In other words, political behavior, whether at the individual or the institutional-level, is driven by incentives. Thus, responses to environmental problems are driven by which political rewards are available. Importantly, political rewards can be both an incentive and disincentive for action, so in some cases, inaction may be the best option. Most incentives are extrinsic in that they provide an external reward that is predicated on a certain behavior, but not inherent in the behavior itself (Kreps, 1997; Park & Word, 2012). But motivations are complex and behaviors tend to be in response to a mixture of different incentives that affect individual decision-making within institutions, with institutional behavior as an aggregation of groups of people who operate within its bounds. While this assumes that policy actors are largely self-rational, it does not assume that self-rational is defined strictly in a traditional economic sense (i.e., dollars and cents), so altruism may be just as important in the calculus of self-rational behavior as more conventional incentives.

So, how are these incentives structured? Let us start with those incentives that tie directly to geographic and institutional jurisdictions, as most elected officials and bureaucrats operate based on a narrowly defined district or service area that bounds both the constituencies that they serve and their control over problems. As most politicians will tell you, the first rule of getting elected is: get re-elected. Or, in other words, electoral incentives are a major driver of the choices that elected officials make. In a most basic sense, elected officials represent their constituencies, so their choices are at least partially tied to public beliefs, attitudes, or opinions on policy issues. That is, they try to align public policy with citizen preferences in order to satisfy their service demands (Maestas, 2000; Lizzeri & Persico, 2001; List & Sturm, 2006; Chang, 2008; Aidt & Shvets, 2012). While this assumes that elected officials serve as a delegate of their constituencies and only act as directed, some argue that elected officials are more like trustees who are empowered to make choices based on their best judgement. Consequently, elected officials may not always choose policies that directly align with public opinion if they believe alternative choices better serve their constituents' interests (Fox & Shotts, 2009; Rehfeld, 2009). In either case, the overarching goal is to create policies that improve or maintain expectations for public services within their districts in order to garner electoral support.

Although bureaucrats are not directly faced with elections, they are subject to citizen demands as well, which incentivizes or disincentivizes their behavior. While classical theory concerning self-rational bureaucrats assumes that they are driven by a desire to maximize power and resources, I prefer to make a milder assumption in that they are driven by a need to ward off threats. That is, some bureaucrats may be interested in creating powerful fiefdoms from which they rule over policy areas, but most simply want to do their jobs without the threat of resource retraction, coercive oversight, or termination (i.e., satisficing) (Simon, 1997; Waterman & Meier, 1998). So, we can largely expect bureaucrats to be responsive to public preferences concerning levels of service in order to preempt criticisms about their efficacy (Waterman & Meier, 1998; Reed, 2014; Fowler, 2019). Certainly, low-performing bureaucratic agencies are subject to a higher degree of public criticism that makes them targets for politicians. On the other hand, high-performing agencies (or at least those that maintain the status quo) have more political capital to negotiate for resources or program adjustments. Consequently, bureaucrats align their decisions and behaviors with citizen preferences in order to satisfy their demands for public services in much the same way as elected officials.

While political rewards may not be a zero-sum game, many rewards are rivalrous, which has precipitated an interjurisdictional competition. Institutions "win" this competition by making decisions that are rewarded with an influx of new citizens, businesses, resources, national attention, and so on. In environmental policy, this has led to two distinct phenomena: a "race-to-the-top" and a "race-to-the-bottom." In the former, policymakers adopt regulations above and beyond other jurisdictions that are then aggressively enforced to improve environmental quality. In turn, environmentally conscious citizens and businesses are drawn to these jurisdictions, which is increasingly common as environmentalism and environmental quality have grown as sociopolitical values and quality of life measures, respectively. In the latter, policymakers create a lax regulatory regime and passive enforcement agenda in order to attract businesses and industries via minimal regulatory compliance costs. While this may lead to lower environmental quality, it also contributes to economic development, job creation, and tax bases, which help fortify re-election bids and bureaucratic security in economically depressed areas (Potoski, 2002; Rabe, 2006; Woods, 2006; Konisky, 2007; Chang, Sigman, & Traub, 2014). While the "races" are commonly viewed from the perspective of subnational governments, they also affect policy choices at the national-level in terms of international competition (Prakash & Potoski, 2006).

Interjurisdictional competition also creates incentives to take advantage of other policy actors. For instance, one state may be willing to implement lax regulations if they know that pollution is likely to drift into a neighboring state, meaning that the environmental risk will be shared between states, but the benefits of a lax regulatory regime will be exclusive (Xepapadeas, 1991, 1992). On the other hand, another state may be unwilling to implement strict air quality regulations if a neighboring state pursues such an aggressive enforcement agenda that regional air quality improves, and allows other states to free-ride on the benefits of a lower concentration of pollutants across a regional airshed (Konisky & Woods, 2010). Furthermore, distribution of authorities in a federal system tends to be driven by credit claiming and blame avoidance behavior, more than efficient or effective policy management. Policy actors typically want to claim credit for popular programs that produce benefits and shift blame to others for unpopular programs that are costly or low performing (Volden, 2005; Weimer, 2006). In general, these types of behavior are driven by attempts to maximize benefits and minimize costs to constituencies by taking advantage of environmental problems that transect jurisdictions.

While political rewards tied to jurisdictions are the biggest impediment to approaching environmental challenges holistically, national political agendas create nonlocalized rewards by inserting outside influence into jurisdictional-based decision-making. More specifically, organized political interest groups, cognizant of pluralistic policy venues, routinely coordinate their efforts across jurisdictions in order to increase the likelihood of successfully achieving their goals. For instance, political parties plan election strategies across multiple districts, and interest groups shop for policymaking venues friendly to their causes (Cox, 1999; Holyoke, Brown, & Henig, 2012; Ley & Weber, 2015). Of course, these efforts bring along with them campaign donations and electoral support that may rival those available from political interests inside of jurisdictions. Additionally, these outside interests lobby bureaucratic agencies to align their implementation efforts with their policy preferences, either directly or indirectly via their influence with elected officials (Waterman, Rouse, & Wright, 1998; Furlong & Kerwin, 2005). In other words, national coordination provides delocalized political rewards to policy actors. On the positive side, this leads some policymakers to think about how environmental policy connects with national issues that transcend a narrow jurisdictional focus. On the negative side, special interests may discourage environmental policies that benefit a specific jurisdiction if it is antithetical to their national agenda.

Finally, not all incentives are extrinsic; some are intrinsic in that they provide a psychological reward that is inherent in the behavior itself. For many who enter the public service, they do so because they believe in a duty to others that is bigger than one's self (Perry, 1990; Broockman, 2013). In general, intrinsic motivators, such as altruism, may account for behaviors that cannot be explained by conventionally defined self-rational behavior. As such, policy actors may be willing to make decisions that provide no extrinsic benefit to themselves or their constituents because they believe in a moral obligation that transcends their narrowly defined institutional responsibilities (Popp, 2001; Rodriguez & Leon, 2004). For example, a politician in a conservative district may be willing to face an electoral backlash to support rigorous environmental regulations if she believes in a moral obligation to protect the environment; or a bureaucrat may be willing to levy a fine against a politically powerful industry because she believes she has an ethical responsibility to do so, despite pressure from elected officials not to. While altruism is an important component of political behavior, few people are solely motivated along these lines. Rather most policy actors respond to a combination of extrinsic rewards tied to strategic political thinking and a set of intrinsic rewards tied to their own feelings of ethical or moral responsibilities (Frey, 1999; Park & Word, 2012).

While some political rewards transcend jurisdictions, they are primarily structured so that policy actors, namely elected officials and bureaucrats, prioritize their narrowly defined jurisdictions when making decisions. It is certainly easy to understand why policymakers in Idaho would have little incentive to adopt policies that benefit people in Maine, so we can hardly expect them to expend the political capital in order to do so. Moreover, there is a large disincentive for even the most altruistic or nationally connected Idaho policymaker to do so if the benefits of a hypothetical policy are distributed in Maine but the costs are borne by those in Idaho. Likewise, this extends to national policy where Congressmen must balance the concerns of their districts against those of the entire nation. Political rewards of any policy choice, such as environmental protection, must be localized to specific geographic or institutional jurisdictions in order to incentivize policy actors to action. Otherwise, policy actors will follow the path of least resistance and uphold the status quo (i.e., inaction) and avoid the path of most resistance, which involves turning their constituents into forced riders (i.e., the inverse of free-riders) that are required to share costs without enjoying benefits.

Public Opinion

A key source of political pressure on policymakers is public opinion. Public opinion represents the collective perspective of a group of citizens within a defined community. It is an aggregation of how individual citizens think about issues or problems and helps us to identify the prevailing collective norms that different societies have constructed around policy issues. Public opinion is generally thought of as a representation of beliefs and attitudes, with those beliefs and attitudes affecting the frame in which people interpret the world around them. In turn, how people interpret the world influences their behavior, including their political behavior as they participate in political decision-making or interact with governmental agencies or other citizens. From a broader perspective, patterns of public opinion, beliefs, attitudes, and behaviors are collectively thought of as culture, and when these are directed at political issues, political culture. Thus, inter-state differences in public opinion on environmental issues is really a manifestation of much larger differences in how citizens of different states interpret environmental issues (Erikson, McIver, & Wright, 1987; Feldman, 1988; Erikson & Tedin, 2016). Of course, there are important correlations between public opinion and demographic factors, such as age, gender, race, income, education, and partisanship, that shape these differences based on state population characteristics (Daniels, Krosnick, Tichy, & Tompson, 2012).

Although it is a highly debated and nuanced topic, scholars identify two causal pathways that connect public opinion on policy issues to policy responses from governmental actors. First, attitudes about environmental issues influence how citizens vote in elections, which leads to elected officials holding similar patterns of beliefs about the environment to their constituents. While some studies indicate that pro-environmental voters are more likely to favor candidates with pro-environmental agendas, other studies find that the environment is not an important enough issue for most voters to motivate a change in candidate preferences (Ladd & Bowman, 1995; Alvarez & Nagler, 1998; Guber, 2001; Davis & Wurth, 2003; Johnson, Brace, & Arceneaux, 2005; Daniels, Krosnick, Tichy, & Tompson, 2012). Second, once in office, elected officials respond to public opinion in order to maintain their electoral support. For instance, previous research finds associations between environmental public opinion and legislative voting behavior, where legislators are more likely to vote in favor of environmental policies when their constituencies hold generally pro-environmental beliefs (Stimson, MacKuen, & Erikson, 1995; Johnson, Brace, & Arceneaux, 2005; Agnone, 2007; Weaver, 2008; Daniels, Krosnick, Tichy, & Tompson, 2012). In lieu of indirectly influencing policy choices through elected officials, public opinion may also directly influence voter behavior on ballot propositions or other direct democracy initiatives (Guber, 2003; Daniels, Krosnick, Tichy, & Tompson, 2012).

While the latent dimensions of public beliefs, attitudes, or culture present within public opinion have been debated by scholars, Fowler (2016) argues that these are a function of perceived responsibilities for the environment, as it relates to the government, the marketplace, and the individual. In other words, if we share collective ownership of the environment, with whom does the responsibility lie for protecting it? From this perspective, public attitudes are then a manifestation of cultural norms surrounding how responsibilities for collective action are shared between public organizations, private enterprises, and individual citizens. In turn, collective understandings of this balance shape public opinion of the environment as citizens interpret public policies and environmental conditions that surround them (Carman, 1998). As such, individual citizens may feel a stronger responsibility to engage in pro-environmental behaviors but do not support environmental regulations, or vice versa. This connects with other conceptualizations of broader political culture that identifies attitudinal patterns that define "good" government in relation to the provision of public goods, preservation of marketplaces, or upholding of social structures (Elazar, 1984). In general, this would suggest that environmental attitudes and beliefs are much more complex than the simplistic pro-environment versus pro-business rhetoric that typically surrounds policy debates.

Along these lines, Fowler (2016) also identifies inter-state differences that indicate public opinion in most states is pro-environmental but not necessarily supportive of government doing more to protect the environment. Notably, respondents in Northeastern states (e.g., Pennsylvania) were more likely to support more government spending on the environment than those in Southern/Southeastern (e.g., Mississippi, South Carolina) or Western states (e.g., Idaho, Montana), while respondents in Western states were more willing to sacrifice their standard of living for environmental protection and to feel that they can have an individual impact on the environment compared to those in

Northeastern and Southeastern states. While respondents in most states tend to be in some agreement with pro-environmental stances, there were a handful of states in the South/Southeast (e.g., Arkansas, Kentucky) in which respondents generally rejected the environment as a political issue. In general, studies find that states with more rural, conservative-leaning populations are less likely to support more government action to protect the environment, as compared to states with more urban or progressive-leaning populations (Daniels, Krosnick, Tichy, & Tompson, 2012; Fowler, 2016). Additionally, while geographic disparities in environmental beliefs are decreasing over time as Americans are becoming more pro-environment, how states rank in terms of support for government spending has largely remained stable (Brace, Arceneaux, Johnson, & Ulbig, 2004; Fowler, 2017; Kim & Urpelainen, 2018).

Map 3.1 presents the geographic distribution of our public opinion index, which is a composite measure from recent studies.[1] States with higher scores have higher proportions of citizens holding pro-environmental opinions. It largely appears that the Southeastern states are least likely to hold pro-environmental public opinion, so we should largely expect there to be little electoral pressure in states such as Arkansas, Tennessee, or South Carolina to become more aggressive with environmental protection. Of course, states such as Georgia and Louisiana appear to be slightly more pro-environment but are still below average in comparison to the rest of the country. On the other hand, Florida appears to be an exception to that regional pattern, which may be influenced by a large progressive population in the southern part of the state that is not as conventionally conservative as the panhandle area. On the other hand, the Northeastern states show a divergent pattern with almost every state being above average in pro-environmental public opinion. Therefore, we

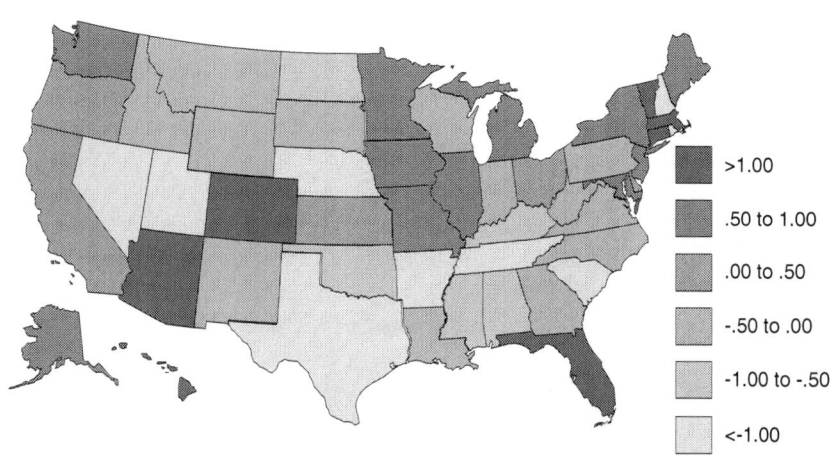

MAP 3.1 Public Opinion Index

should certainly expect there to be tremendous pressure on policymakers in states like Connecticut, Massachusetts, and Vermont to protect the environment. Interestingly though, New Hampshire sits amidst these states and has the lowest level of support in the region.

As we move west, a much higher degree of inconsistency across regions emerges, which may reflect geographic, environmental, and demographic diversity in the Midwest and Western regions. For instance, there appears to be above average pro-environmental public opinion among the Great Lakes states (e.g., Michigan, Minnesota, Illinois), while this level of support begins to dissipate along the Great Plains with states like Nebraska, Oklahoma, and Texas having relatively low support for the environment. In general, the Western states seem to be above average in their support of the environment. The glaring exceptions are Nevada and Utah, which have among the lowest pro-environmental support in the nation. This may be explained by both geographic and demographic factors when compared to neighboring states, such as Arizona and Colorado, that have much higher levels of support as well as different environmental amenities and population characteristics. In general, this level of diversity would suggest that public opinion on the environment is quite complex and influenced by numerous factors internal to states, so we can expect a divergence in state responses to environmental problems as policymakers respond to their constituents.

Environmental Advocacy and Interest Groups

The second key source of the political pressure on policymakers is environmental advocacy groups or organized interests. In general, advocacy groups create organization around political interests and use that organization to mobilize financial and political resources to influence policy choices. This is particularly important for issues like the environment in which there is collective ownership by the public but that ownership interest is not otherwise formally represented. Environmental advocacy groups emerged as early as the late 19th century to provide formal representation to public interest in protecting and conserving natural resources. While groups like the Sierra Club and the National Audubon Society have existed for over a century, most contemporary groups, such as the Natural Resources Defense Council and Greenpeace, find their origins in the 1960s and 1970s as a result of the environmental movement. As such, many of these groups position themselves as the modern defenders of that legacy, even though there have been important questions about their efficacy (Shellenberger & Nordhaus, 2009; Duffy, 2012). Advocacy groups may be most visible in public outreach campaigns, but their participation in the political process takes many forms, including campaign contributions, organizing political rallies or events, and direct lobbying of policymakers. In general, advocacy groups create political pressure during three phases of the policy process: agenda-setting, decision-making, and implementation (Andrews & Edwards, 2004).

First, given that policymakers only consider a finite number of issues at any time, advocacy groups compete to see their issues make it onto the political agenda through public outreach or education campaigns, the use of mass media, or direct lobbying of policymakers. The goal here is to turn public attention towards environmental issues, and many scholars argue that advocacy groups are most effective during this phase (Andrews & Edwards, 2004; Pralle, 2006). Second, advocacy groups work to influence how policymakers decide on specific issues by creating political pressure that directs them towards particular choices. In some cases, this may come in the form of electoral support (i.e., campaign contributions); in other cases, it involves testifying in legislative hearings or producing research reports in support of policies (Andrews & Edwards, 2004; Grossman, 2006; Witko, 2013). Third, advocacy groups also attempt to influence how policies are put into practice by lobbying administrative agencies on their interpretation of legislation, during the rulemaking process, or on how policies are monitored and evaluated. This is most commonly achieved through leveraging political connections with elected officials (i.e., iron triangles) or direct lobbying of administrative officials, but may also involve drawing unwanted public attention to administrative agencies (e.g., protests) (Andrews & Edwards, 2004; Yackee, 2006; Rinfret, 2011; Fowler, 2019)

In most cases, the power of advocacy groups is a function of their membership and financial resources (Andrews & Edwards, 2004). In terms of the former, advocacy groups are more powerful if they can mobilize a significant number of citizens to pressure policymakers, which may occur through both passive (e.g., letter-writing campaigns) and active (e.g., protests) strategies (Oberholzer-Gee & Waldfogel, 2005). Of course, environmental advocacy groups are also in competition with industry groups (or other interests) pursuing contradictory goals, so it is necessary that they have access to a membership base that can provide them with the political capital necessary to access policymakers and influence policy. To that end, if membership in an environmental group is relatively common within a state, environmental interests have potentially more influence, as elected officials seek out endorsements or access to membership information that can be used in fundraising or voter outreach. On the other hand, if membership is relatively uncommon, elected officials may be unconcerned with the salience of the environment as a political issue (Nownes & Freeman, 1998). Previous studies of environmental groups indicate higher membership rates in the West and Northeast, and particularly low membership rates in the South/Southeast (Wikle, 1995; Mazur & Welch, 1999).

Advocacy groups can also amass influence and political power by directing their financial resources towards candidates running for elected offices, either directly through contributions to candidate campaigns or indirectly through their own campaign activities. This is a key mechanism by which advocacy groups can build influence from outside jurisdictions. In other words, they may donate to candidates in jurisdictions in which the environment is not a particularly salient topic in order to encourage elected officials to add environmental issues onto the political agenda (Holman & Claybrook, 2004; Witko, 2013). In recent years, advocacy groups have become particularly sophisticated at strategically venue-shopping and making campaign contributions (Pralle, 2010; Ley & Weber, 2015). Consequently, even if there is low public support or group membership within a state, national environmental groups can direct campaign contributions to candidates who may be in search of electoral support. Unsurprisingly, data from the National Institute on Money in Politics (NIMP) indicates that campaign contributions from pro-environmental groups make up a larger portion of total contributions in Western states than in any other region. Interestingly, there are also several Southeastern states (e.g., North Carolina, Virginia) in which environmental groups spend big as well (NIMP, 2019).

Map 3.2 presents the geographic distribution of our advocacy group index.[2] States with higher scores have higher rates of environmental group membership and campaign contributions from environmental advocacy groups. There are few clear regional patterns that emerge from the map. The Western states appear to be a hotspot for environmental advocacy groups, with a majority of states having the highest level of activity (e.g., California, Colorado, Montana). This is not particularly shocking given the political culture and environmental amenities in those states. There are a few exceptions, namely Arizona, Utah, and Wyoming in the Rocky Mountain region, although those states are only slightly below average. On the other hand, the South/Southeast appears to be the polar opposite of the West, with very little environmental advocacy group activity. Alabama is an interesting exception; further examination indicates this

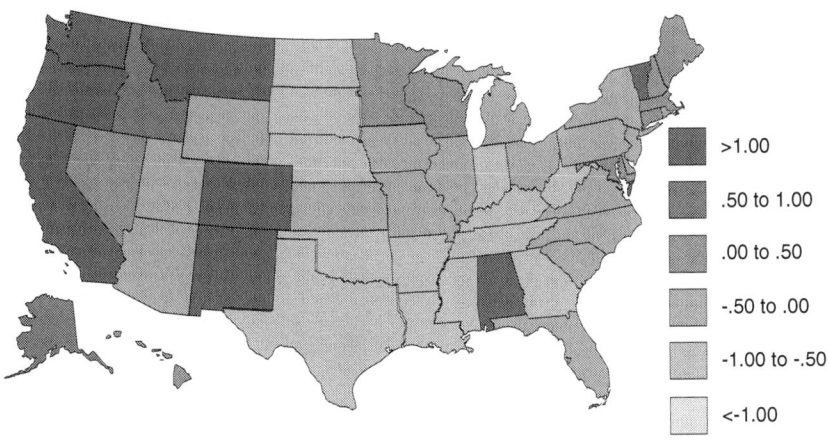

▉	>1.00
▓	.50 to 1.00
▒	.00 to .50
▒	-.50 to .00
░	-1.00 to -.50
░	<-1.00

MAP 3.2 Advocacy Index

score is driven mostly by campaign contributions. Notably, Alabama has among the lowest environmental spending and highest rates of pollution in the region, so environmental groups may specifically target it for those reasons. In general, it appears that environmental groups are less influential in states with large rural populations and where environmental issues are relatively unimportant to the populace. This would also explain why the Great Plains states also experience a relative dearth of environmental group activity. Finally, there appears to be a fairly moderate, consistent level of activity across most of the Midwest and Northeast.

Comparative Policy

A final source of political pressure on policymakers is how states compare to their neighbors, which is indicative of both how states respond to horizontal competition and regional sociopolitical norms. More specifically, our concern here is how states are different from their neighbors, which provides us with some additional insight into understanding which individual states may be trying to gain an advantage over their neighbors or challenge regional norms. In other words, we are concerned with whether states are providing more or less environmental protection than their neighbors. If more, we can assume there is more political will for environmental protection in those states, even if that is not indicated by public opinion or environmental group activities. Environmental protection, like any other social activity, is subject to the creation and maintenance of social norms; that is, norms surrounding what levels of pollution or government intervention are acceptable (Stern, Dietz, & Black, 1985; Berry & Berry, 2014). In general, we can assume that levels of pollution or government spending are normally distributed across states, meaning that most observations are concentrated around the average and there are few outliers with either very high or very low levels of pollution or spending. For instance, 28 states spend between 1% and 2% of their annual budgets on the environment, and 30 states spend between $50 and $100 per capita on the environment (Census, 2019).

It is not a coincidence that a majority of states have similar spending patterns. Rather, it is a function of what people believe is a "normal" amount to spend on the environment and/or what it costs to provide a level of environmental protection that people believe is "normal." Certainly, those norms have evolved over time, but there is a fairly high degree of consistency in spending patterns across states over the last few decades. Therefore, the concept of "norms" of environmental protection is an important aspect of how policy actors perceive government action as they consider what level of environmental protection should be provided. In other words, their answer to that question is dependent on what their neighbors are doing, as well as what has been done in the past. While these norms are influenced by national trends to a degree, closer states (i.e., neighboring states) tend to have more influence. Looking at neighboring states provides policymakers with signals for what is and is not acceptable, or at least provides them a frame of reference to benchmark their own policies (Baybeck, Berry, & Siegel, 2011). As our discussion of public opinion and advocacy groups indicates, there are regional patterns of environmental politics, so norms in one part of the country may be different than in other parts.

Furthermore, many scholars view this from the lens of competition in which states compete with each other to provide a "package" of services that best matches the preferences of self-rational political actors seeking to maximize their benefits. In environmental policy, this "package" consists of environmental quality on one hand and the extent of government regulations on the other. To this end, some policymakers may determine that providing better environmental quality than their neighbors is a way to increase the value of public services in their states and make it more attractive to environmentally conscious citizens, while other policymakers may determine that reducing regulations is a way to decrease barriers to economic activity or administrative costs in order to spur economic development. Consequently, there are two forms of competition that exist, with some states finding value in providing the lowest level of environmental protection possible, and others, value in providing the highest level possible (Potoski, 2002; Levinson, 2003; Holzinger & Knill, 2004; Wilson & Damania, 2005; Prakash & Potoski, 2006; Rabe, 2006; Woods, 2006; Konisky, 2007). In either case though, those levels are benchmarked against neighboring states, so the marginal utility in providing more or less protection is reduced as states diverge from their neighbors. In other words, for a state that already has better environmental quality or fewer regulations than its neighbors, there is little additional incentive to continue to improve quality or reduce regulations, respectively.

Additionally, causes and consequences of environmental problems are not always concentrated within a single state, which creates opportunities for states to take advantage of their neighbors (Feiock & Scholz, 2009; Konisky & Woods, 2010). More specifically, state policymakers may find themselves facing policy choices that impact not only environmental quality in their state, but also in neighboring states. In this case, policymakers may see the best option in one of three ways: 1) be more aggressive than your neighbors to offset any failures on their part; 2) be less aggressive than your neighbors so you can free-ride on their efforts; or 3) match your level of service with your neighbors' so you are contributing to the solution but not taking on more burden than others. Certainly, the decision calculus is telling of the political rewards for environmental protection. If policymakers are determined to improve environmental quality, they will likely choose the first option and work to that end regardless of how their neighbors respond. On the other hand, if policymakers are disinterested in the environment, they will likely choose the last option and take advantage of the situation. Still other policymakers may find just enough

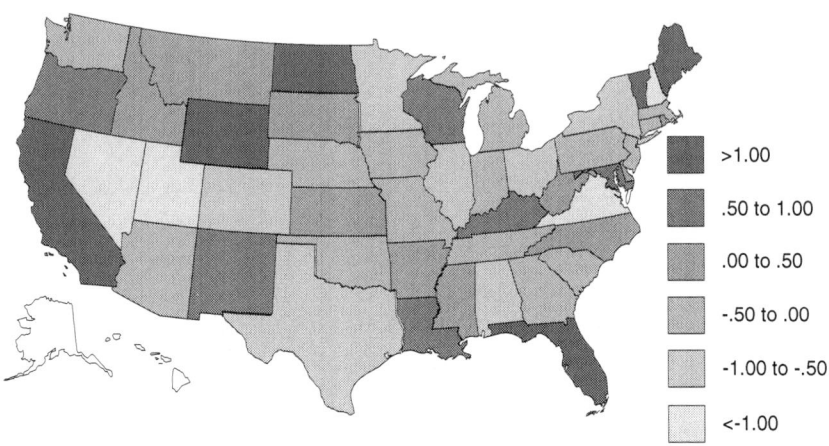

MAP 3.3 Comparative Policy Index

political incentive to not be seen as shirking their environmental responsibilities, but not enough to make unilateral investments to improve environmental conditions. Consequently, how states compare to their neighbors is indicative of their political will to protect the environment.

Map 3.3 presents the geographic distribution of our comparative policy index.[3] States with higher scores have fewer toxic releases and higher environmental expenditures than their neighbors. Unlike the other indices presented here, this index is specifically designed to identify states in comparison to their neighbors, rather than to identify regional patterns. Therefore, there are few general observations about geographical distributions to make; however, we can identify states that are relative leaders or laggards. In terms of the former, Maine and Vermont in the Northeast, Florida in the Southeast, North Dakota in the Midwest, and California and Wyoming in the West are outpacing their neighbors in environmental protection. In terms of the latter, New Hampshire in the Northeast, Virginia in the Southeast, and Nevada and Utah in the West have all fallen behind their neighbors. Although not in the lowest grouping, Illinois, Ohio, and Minnesota in the Midwest fall into the second to last grouping of states. Equally as notable are states whose comparative policy index hovers around the average of their neighbors, such as Connecticut and Rhode Island in the Northeast, Georgia and Mississippi in the Southeast, Indiana and Kansas in the Midwest, and Montana and Washington in the West.

Political Incentives Index

As the discussion of public opinion, advocacy groups, and comparative policies indicate, there is a lot of variation in the political context from which states manage the environment, and those facets reveal different sources of pressures to be more or less aggressive. Given this, I created the political incentives index[4] (see Map 3.4) by combining these measures, the goal of which is to further understand which states are more amenable to environmental protection. This map shows many of the same regional patterns that our previous maps showed: a concentration of pro-environmental support in the West, a dearth of support in the South/Southeast, and moderate levels throughout the Northeast and Midwest. Of course, there are exceptions to those regional generalizations, with states like Nevada and Utah serving as regional outliers. Looking at individual states, those with the highest scores were exclusively in the West (i.e., California, Hawaii, Oregon, and Wyoming) or the Northeast (i.e., Maine, Massachusetts, and Vermont). On the other hand, states with the lowest scores showed much more regional variation: New Hampshire in the Northeast; Arkansas, Tennessee, and South Carolina in the South/Southeast; Nebraska in the Midwest; and Nevada and Utah in the West.

The political incentives index provides us insights into how states differ in their political will to commit to environmental protection, or how they determine what the appropriate level of environmental protection is. Certainly, for states at the higher end of this scale, there is an expectation that environmental protection be aggressive, rigorous, and prioritized above competing demands; for states at the lower end, there is likely an expectation that environmental protection be minimal with the government doing only what is necessary to guarantee human health or meet externally mandated requirements. These polarized perspectives highlight both the pros and cons of decentralization as a core tenet of environmental federalism. On one hand, it means that citizens in each state have influence over the level of environmental protection they believe is appropriate for their communities; on the other hand, it means that some citizens are exposed to much more pollution than others, drawing

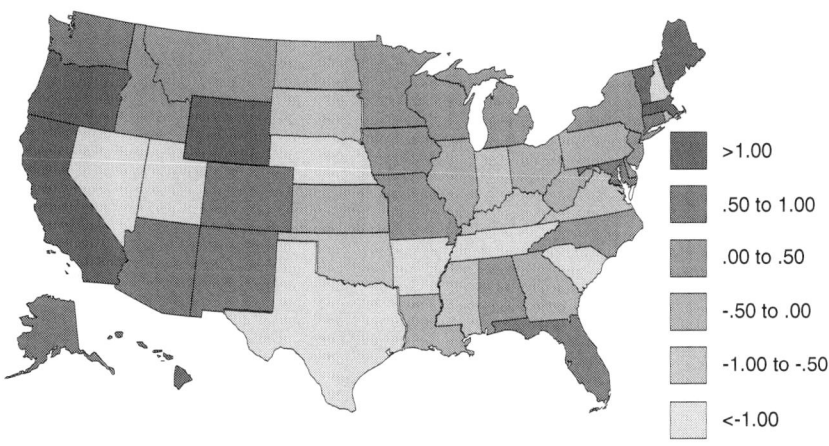

> .00
.50 to 1.00
.00 to .50
-.50 to .00
-1.00 to -.50
<-1.00

MAP 3.4 Pro-Environmental Politics Index

questions of equity. Equally as important are the states in the middle, which may be pulled in competing directions as there is not enough pressure to make the environment a policy priority nor to ignore it. The political incentives index provides us one of two dimensions to understand how states play their role as linchpins in environmental federalism. The other dimension is administrative capacities, which we will examine in the next chapter.

Notes

1. I pulled state-level public opinion estimates from Fowler (2016) on three survey items and Kim and Urpelainen (2018) on one survey item. I then calculated z-scores to create a common scale, and combined the four items into a composite measure using a weighted average that double counted the survey items on government spending from both studies. While I believe other elements of environmental attitudes are important, the government spending items are most directly related to public interest in the government's role in protecting the environment, so I weighted those more heavily. Fowler (2016) and Kim and Urpelainen (2018) use similar methodologies and data: Multilevel Regression and Postestimation (MPR). MRP follows a two-step process where a multi-level regression model is estimated based on a series of demographic variables at level-1 and states at level-2. Then, public opinion estimates for each state are calculated based on the unique state-level intercept and population demographics. Studies show this technique to be more accurate at estimating state-level public opinion using national surveys than more conventional aggregation techniques (Pacheco, 2011; Warshaw & Rodden, 2012). Both studies use data from the General Social Survey. Examining data from 2000 to 2010, Fowler (2016) uses three questions: are we spending too much, too little, or about the right amount on improving and protecting the environment?; is it just too difficult for someone like me to do much about the environment?; and, how willing would you be to accept cuts in your standard of living in order to protect the environment? On the other hand, Kim and Urpelainen (2018) use the same government spending item but examine data from 1973 to 2012, providing a more robust estimation model. For the government spending item, findings were still highly correlated ($r = .75$).

2. I created a composite measure of environmental group membership and campaign contributions from pro-environmental groups. First, I calculated the average rate of Sierra Club members between 2000 and 2010. Although Sierra Club is but one environmental group, it has been previously used as an indicator for prevalence of environmental group membership at the state-level (Konisky & Woods, 2012). Additionally, it is highly correlated, with more robust estimations that incorporate multiple groups, such as Mazur and Welch (1999) ($r = .83$). Second, I pulled data from the NIMP's FollowtheMoney.org on campaign contributions to gubernatorial, state senate, state house, and other statewide office candidates between 2000 and 2010 (NIMP, 2019). Then, I calculated the portion of those donations from pro-environmental groups. Correlation indicated a strong relationship between group membership and campaign contributions ($r = .32$). Third, I calculated z-scores to create a common scale for both variables and combined into a composite measure. Although there are potentially other indicators, I believe these two get at the core of state-level environmental group strength.

3. I created a composite measure that compares states to their neighbors based on average toxic releases per capita, environmental expenditures per capita, and environmental expenditures as a percentage of total state expenditures from 2000 to 2010. I used data from the EPA (2019) and US Census Bureau (2019). First, I calculated the neighbor average using adjacent states for all three variables for every state, and I took the difference

between the state observation and neighbor average. In order to create consistency in direction related to pro-environmental commitment, I subtracted the state observation from the neighbor average for toxic releases, but use the inverse formula for expenditure variables. Consequently, less toxic releases and more expenditures magnify rather than offset each other, which is consistent with our measurement goals of identifying states that are ahead or behind their neighbors in terms of pro-environmental policy behaviors. Second, I calculated z-scores to create a common scale for all three variables and combined into a composite measure. Since Alaska and Hawaii have no neighbors, they were both scored zero on this index. Correlations between expenditure variables were high ($r = .88$), but were very weak with toxic releases ($r = .15, .04$). Bacot and Dawes (1997) indicates there is not a direct relationship between environmental spending on pollution levels, so this weak correlation is not surprising. However, I believe these different variables provide some insights into both how states' political processes ascribe value to environmental protection and whether that has an impact on the environment, both of which are of interest to us as we compare political incentives for environmental protection.

4. Since all three measures were already measured on a common scale via z-scores, I combined into a composite measure by calculating the average for each state. While the public opinion index is strongly correlated with both advocacy ($r = .30$) and comparative ($r = .22$) indexes, there is a weaker relationship between advocacy and comparative indexes ($r = .11$). Given that I designed the comparative index to capture a different dimension than the other two in order to isolate individual states that stand out from their region, I am not surprised that it has a weak correlation, because the other two indexes capture regional patterns.

References

Agnone, J.M. 2007. Amplifying Public Opinion: The Policy Impact of U.S. Environmental Movement. *Social Forces* 85(4): 1593–1620.

Aidt, T.S. & J. Shvets. 2012. Distributive Politics and Electoral Incentives: Evidence from Seven U.S. State Legislatures. *American Economic Journal* 4(3): 1–29.

Alvarez, R.M. & J. Nagler. 1998. Economics, Entitlements, and Social Values: Voter Choice in the 1996 Presidential Election. *American Journal of Political Science* 42: 1349–1363.

Andrews, K.T. & B. Edwards. 2004. Advocacy Organizations in the U.S. Political Process. *Annual Review of Sociology* 30: 479–506.

Bacot, A.H. & R.A. Dawes. 1997. States Expenditures and Policy Outcomes in Environmental Program Management. *Policy Studies Journal* 25(3): 355–370.

Baybeck, B., W.D. Berry, & D.A. Siegel. 2011. A Strategic Theory of Policy Diffusion via Intergovernmental Competition. *Journal of Politics* 73(1): 232–247.

Berry, F.S. & W.D. Berry. 2014. Innovation and Diffusion Models in Policy Research. In *Theories of the Policy Process*, 3rd ed., edited by P.A. Sabatier & C.M. Weible (pgs. 307–359). Boulder, CO: Westview Press.

Brace, P., K. Arceneaux, M. Johnson, & S.G. Ulbig. 2004. Does State Political Ideology Change over Time? *Political Research Quarterly* 57(4): 529–540.

Broockman, D.E. 2013. Black Politicians Are More Intrinsically Motivated to Advanced Blacks' Interests: A Field Experiment Manipulating Political Incentives. *American Journal of Political Science* 57(3): 521–536.

Carman, C.J. 1998. Dimensions of Environmental Policy Support in the United States. *Social Science Quarterly* 79(4): 717–733.

Chang, E.C.C. 2008. Electoral Incentives and Budgetary Spending: Rethinking the Role of Political Institutions. *Journal of Politics* 70(4): 1086–1097.

Chang, H.F., H. Sigman, & L.G. Traub. 2014. Endogenous Decentralization in Federal Environmental Policies. *International Review of Law & Economics* 37: 39–50.

Cox, G. 1999. Electoral Rules and Electoral Coordination. *Annual Review of Political Science* 2:145–161.

Daniels, D.P., J.A. Krosnick, M.P. Tichy, & T. Tompson. 2012. Public Opinion on Environmental Policy in the United States. In *The Oxford Handbook of U.S. Environmental Policy*, edited by M.E. Kraft & S. Kamieniecki (pgs. 461–486). Oxford: Oxford University Press.

Davis, F.L. & A.H. Wurth. 2003. Voting Preferences and the Environment in the American Electorate: The Discussion Extended. *Society & Natural Resources* 16(8): 729–740.

Duffy, R.J. 2012. Organized Interests and Environmental Policy. In *The Oxford Handbook of U.S. Environmental Policy*, edited by M.E. Kraft & S. Kamieniecki (pgs. 504–524). Oxford: Oxford University Press.

Elazar, D. 1984. *American Federalism: A View from the States*, 3rd ed. New York: Harper Row.

Erikson, R.S., J.P. McIver, & G.C. Wright. 1987. State Political Culture and Public Opinion. *American Political Science Review* 81(3): 797–813.

Erikson R.S. & K.L. Tedin. 2016. *American Public Opinion*, 9th ed. New York: Routledge.

Feiock, R.C. & J.T. Scholz. 2009. Self-Organizing Governance of Institutional Collective Action Dilemmas: An Overview. In *Self-Organizing Federalism: Collective Mechanisms to Mitigate Institutional Collective Action Dilemmas*, edited by R.C. Feiock & J.T. Scholz (pgs. 3–32). Cambridge: Cambridge University Press.

Feldman, S. 1988. Structure and Consistency in Public Opinion: The Role of Core Beliefs and Values. *American Journal of Political Science* 32(2): 416–440.

Fowler, L. 2016. The States of Public Opinion on the Environment. *Environmental Politics* 25(2): 315–337.

Fowler, L. 2017. Tracking State Trends in Environmental Public Opinion. *Social Science Journal* 54(3): 287–294.

Fowler, L. 2019. Problems, Politics, and Policy Streams in Policy Implementation. *Governance* 32(3): 403–420.

Fox, J. & K.W. Shotts. 2009. Delegates or Trustees? A Theory of Political Accountability. *Journal of Politics* 71(4): 1225–1237.

Frey, B.S. 1999. Morality and Rationality in Environmental Policy. *Journal of Consumer Policy* 22(4): 395–417.

Furlong, S.R. & C.M. Kerwin. 2005. Interest Group Participation in Rule Making: A Decade of Change. *Journal of Public Administration Research & Theory* 15(3): 353–370.

Grossman, M. 2006. Environmental Advocacy in Washington: A Comparison with Other Interest Groups. *Environmental Politics* 15(4): 628–638.

Guber, D.L. 2001. Voting Preferences and the Environment in the American Electorate. *Society & Natural Resources* 14(6): 455–469.

Guber, D.L. 2003. *The Grassroots of a Green Revolution: Polling America on the Environment*. Cambridge, MA: MIT Press.

Holman, C. & J. Claybrook. 2004. Outside Groups in the New Campaign Finance Environment: The Meaning of BCRA and the McConnel Decision. *Yale Law & Policy Review* 22(2):235–259.

Holyoke, T., H. Brown, & J. Henig. 2012. Shopping in the Political Arena: Strategic State and Local Venue Selection by Advocates. *State & Local Government Review* 44(1): 9–20.

Holzinger, K. & C. Knill. 2004. Competition and Cooperation in Environmental Policy: Individual and Interaction Effects. *Journal of Public Policy* 24(1): 25–47.

Johnson, M., P. Brace, & K. Arceneaux. 2005. Public Opinion and Dynamic Representation in the American States: The Case of Environmental Attitudes. *Social Science Quarterly* 86(1): 87–108.

Kim, S.E. & J. Urpelainen. 2018. Environmental Public Opinion in the U.S. States, 1973–2012. *Environmental Politics* 27(1): 89–114.

Konisky, D.M. 2007. Regulatory Competition and Environmental Enforcement: Is There Is a Race to the Bottom? *American Journal of Political Science* 51(4): 853–872.

Konisky, D.M. & N.D. Woods. 2010. Exporting Air Pollution? Regulatory Enforcement and Environmental Free Riding in the United States. *Political Research Quarterly* 63(4): 771–782.

Konisky, D.M. & N.D. Woods. 2012. Measuring State Environmental Policy. *Review of Policy Research* 29(4): 544–569.

Kreps, D.M. 1997. Intrinsic Motivation and Extrinsic Incentives. *American Economic Review* 87(2): 359–364.

Ladd, E.C. & K.H. Bowman. 1995. *Attitudes toward the Environment: Twenty-Five Years after Earth Day*. American Enterprise Institute for Public Policy Research. Washington, DC: AEI Press.

Lasswell, H. 1936. *Politics: Who Gets What, When, How*. New York: McGraw-Hill.

Lester, J.P. 1995. Federalism and State Environmental Policy. In *Environmental Politics and Policy: Theories and Evidence*, 2nd ed., edited by J.P. Lester (pgs. 39–60). Durham, NC: Duke University Press.

Levinson, A. 2003. Environmental Regulatory Competition: A Status Report and Some New Evidence. *National Tax Journal* 56(1): 91–106.

Ley, A.J. & E.P. Weber. 2015. The Adaptive Venue Shopping Framework: How Emergent Group Choose Environmental Policymaking Venues. *Environmental Politics* 24(5): 703–722.

List, J.A. & D.M. Sturm. 2006. How Elections Matter: Theory and Evidence from Environmental Policy. *Quarterly Journal of Economics* 121(4): 1249–1281.

Lizzeri, A. & N. Persico. 2001. The Provision of Public Goods under Alternative Electoral Incentives. *American Economic Review* 91(1): 225–239.

Maestas, C. 2000. Professional Legislatures and Ambitious Politicians: Policy Responsiveness of State Institutions. *Legislative Studies Quarterly* 25(4): 663–690.

Mazur, A. & E. Welch. 1999. The Geography of American Environmentalism. *Environmental Science & Policy* 2(4–5): 389–396.

National Institute on Money in Politics (NIMP). 2019. *Follow the Money*. The Campaign Finance Institute. Available at www.followthemoney.org/

Nownes, A.J. & P. Freeman. 1998. Interest Group Activity in the States. *Journal of Politics* 60(1): 86–112.

Oberholzer-Gee, F. & J. Waldfogel. 2005. Strength in Numbers: Group Size and Political Mobilization. *Journal of Law & Economics* 48(1): 73–91.

Pacheco, J. 2011. Using National Surveys to Measure Dynamic U.S. State Public Opinion: A Guideline for Scholars and an Application. *State Politics & Policy Quarterly* 11(4): 415–439.

Park, S.M. & J. Word. 2012. Driven to Service: Intrinsic and Extrinsic Motivation for Public and Nonprofit Managers. *Public Personnel Management* 41(4): 705–734.

Perry, J.L. & L.R. Wise. 1990. The Motivational Bases of Public Service. *Public Administration Review* 50(3): 367–373.

Popp, D. 2001. Altruism and the Demand for Environmental Quality. *Land Economics* 77(3): 335–345.

Potoski, M. 2002. Clean Air Federalism: Do States Race to the Bottom? *Public Administration Review* 61(3): 335–343.

Prakash, A. & M. Potoski. 2006. Racing to the Bottom? Trade, Environmental Governance, and ISO 14001. *American Journal of Political Science* 50(2): 350–364.

Pralle, S.B. 2006. *Branching Out, Diggin In: Environmental Advocacy and Agenda Setting.* Washington, DC: Georgetown University Press.

Pralle, S.B. 2010. Shopping Around: Environmental Organizations and the Search for Policy Venues. In *Advocacy Organizations and Collected Action*, edited by A. Prakash & M.K. Gugerty (pgs. 177–202). Cambridge: Cambridge University Press.

Rabe, B. 2006. Race to the Top: The Expanding Role of U.S. State Renewable Portfolio Standards. *Sustainable Development Law & Policy* 8(3): 10–17.

Reed, S. 2014. *Building the Federal Schoolhouse.* Oxford: Oxford University Press.

Rehfeld, A. 2009. Representation Rethought: On Trustees, Delegates, and Gyroscopes in the Study of Political Representation and Democracy. *American Political Science Review* 103(2): 214–230.

Rinfret, S.R. 2011. Frames of Influence: U.S. Environmental Rulemaking Case Studies. *Review of Policy Research* 28(3): 231–246.

Rodriguez, M.X.V. & C.J. Leon. 2004. Altruism and the Economic Values of Environmental and Social Policies. *Environmental & Resource Economics* 28(2): 233–249.

Shellenberger, M. & T. Nordhaus. 2009. The Death of Environmentalism: Global Warming Politics in a Post-Environment World. *Geopolitics, History, & International Relations* 1(1): 121–163.

Simon, H.A. 1997. *Administrative Behavior*, 4th ed. New York: Free Press.

Stern, P.C., T. Dietz, & J.S. Black. 1985. Support for Environmental Protection: The Role of Moral Norms. *Population & Environment* 8(3–4): 204–222.

Stimson, J.A., M.B. Mackuen, & R.S. Erikson. 1995. Dynamic Representation. *American Political Science Review* 89(3): 543–565.

U.S. Census. 2019. *Statistical Abstract Series* [online]. Available at www.census.gov/library/publications/time-series/statistical_abstracts.html [Retrieved January 1, 2019].

US Environmental Protection Agency (EPA). 2019. *TRI Basic Data Files: Calendar Year 1987–2017* [online]. Available at www.epa.gov/toxics-release-inventory-tri-program/tri-basic-data-files-calendar-years-1987-2017 [Retrieved January 1, 2019].

Volden, C. 2005. Intergovernmental Political Competition in American Federalism. *American Journal of Political Science* 49(2): 327–342.

Warshaw, C. & J. Rodden. 2012. How Should We Measure District-level Public Opinion on Individual Issues? *Journal of Politics* 74(1): 203–219.

Waterman, R.W. & K.J. Meier. 1998. Principal-Agent Models: An Expansion? *Journal of Public Administration Research & Theory* 8(2): 173–202.

Waterman, R.W., A. Rouse, & R. Wright. 1998. The Venues of Influence: A New Theory of Political Control of the Bureaucracy. *Journal of Public Administration Research & Theory* 8(1): 13–38.

Weaver, A.A. 2008. Does Protest Behavior Mediate the Effects of Public Opinion on National Environmental Policies? *International Journal of Sociology* 38(3): 108–125.

Weimer, D.L. 2006. The Puzzle of Private Rulemaking: Expertise, Flexibility, and Blame Avoidance in U.S. Regulation. *Public Administration Review* 66(4): 569–582.

Wikle, T.A. 1995. Geographical Patterns of Membership in U.S. Environmental Organizations. *Professional Geographer* 47(1): 41–48.

Wilson, J.K. & R. Damania. 2005. Corruption, Political Competition, and Environmental Policy. *Journal of Environmental Economics & Management* 49(3): 516–535.

Witko, C. 2013. When Does Money Buy Votes?: Campaign Contributions and Policymaking. In *New Directions in Interest Group Politics*, edited by M. Grossman. New York: Routledge.

Woods, N.D. 2006. Interstate Competition and Environmental Regulation: A Test of the Race-to-the-Bottom Thesis. *Social Science Quarterly* 87(1): 174–189.

Xepapadeas, A.P. 1991. Environmental Policy under Imperfect Information: Incentives and Moral Hazard. *Journal of Environmental Economics & Management* 20(2): 113–126.

Xepapadeas, A.P. 1992. Environmental Policy Design and Dynamic Nonpoint-Source Pollution. *Journal of Environmental Economics & Management* 23(1): 22–39.

Yackee, S.W. 2006. Sweet-Talking the Fourth Branch: The Influence of Interest Group Comments on Federal Agency Rulemaking. *Journal of Public Administration Research & Theory* 16(1): 103–124.

4

ADMINISTRATIVE CHALLENGES AND THE LIMITS OF ENVIRONMENTAL POLICY IN PRACTICE

On the other side of our matrix for understanding state behavior in environmental federalism are the administrative capacities necessary to operationalize environmental protection. In some states, administrative capacities sufficiently exist to make and implement effective policies that protect the environment, while in other states, administrative agencies struggle to devise appropriate policies or procedures, understand the sources of pollution, and/or enforce regulatory compliance. So, how do we know to what degree administrative capacities exist within individual states? Previous scholars have considered this question within the larger context of state institutional capacity, and how it helps explain patterns of policymaking and implementation (e.g., Bowman & Kearney, 1988). While there are numerous potential ways to answer this question, in this chapter, we look at three dimensions: policymaking, managing information, and creating accountability. Specifically, these dimensions provide insights into the ability of states to develop policies that meet their unique jurisdictional needs, collect and manage data on environmental conditions, and enforce compliance with policies. Collectively, they are indicative of the capacities which can be operationalized to support environmental policy implementation and provide a basis to make inter-state comparisons. The goal of examining these facets is again to determine in which states we may expect to find high degrees of capacity for environmental protection, and in which states low degrees.

Administrative Capacities

Let us first broadly consider administrative capacities for environmental protection. If we assume that sufficient political incentives exist to motivate policy actors to want a high degree of environmental protection, what can we realistically hope to achieve? Even if completely eliminating pollutants from our environment (or halting climate change) were scientifically or technically possible, it would require a tremendous amount of investment of resources to accomplish. Given that we live in a world of scarce resources that have to be divided among competing public interests (i.e., healthcare, welfare, education, law enforcement, transportation, emergency services, and so on), it is unrealistic for us to assume that a level of complete pollution prevention and mitigation over the long-term is practical. Instead, we must consider what realistic capacities exist and how those can be operationalized into environmental protection. Notably, state capacity was a core concern at the EPA in the early days of program primacy. According to former Administrator William Rucklehaus, "EPA had to be sure the states had adequate bureaucratic mechanisms in place before delegating to them the operation and administration of new programs" (EPA, 1993). In other words, there was concern then (and still today) that states lack the ability to make programs work in practice.

So, what do we mean by capacity? Ingraham, Joyce, and Donahue (2003) define it as the "ability to marshal, develop, direct, and control its financial, human, physical, and information resources" (p. 15). In the most basic sense, capacity is the ability to accomplish some kind of task. These tasks are usually defined by organizational missions, so capacity incorporates numerous factors that affect the ability of organizations to do things associated with political or administration processes. For instance, if two agencies have similar budgets or workforces and one agency accomplishes more, the difference between the agencies is their capacity to put their budgets or workforces to good use. Scholarship is rather disjointed on this topic, and there are few clear indications of which capacities are most important or most directly connected to certain types of tasks. In most cases, scholars develop narrow perspectives to understand specific dimensions. In general, capacities can be seen as a function of organization and processes that dictates how resources are utilized. These resources can be diverse and may include workforces, political or social capital, finances, or physical infrastructure (Christensen & Gazley, 2008).

In one of the most comprehensive studies on state-level institutional capacities (and possibly the most citied), Bowman and Kearney (1988) define capacity as the ability of state executive and legislative branches "to respond effectively to change, make decisions efficiently, effectively, and responsively, and manage conflict" (p. 346). Their findings suggest that more conventional measures of institutional capacity such as gubernatorial power, legislative professionalism, staffing, and spending are present, but that other factors such as accountability, information management, and representation are also important. Additionally, their findings indicate that capacities fluctuate widely across states, and subsequent scholars have used their measures to explain outputs across a range of policy areas, including environmental protection (e.g., Travis, Morris, &

Morris, 2004; Fowler, 2016). They also note that capacity should be measured as it relates to specific policy outputs, and there are few "one-size-fits-all" measures that capture the complexity of state capabilities. However, Bowman and Kearney (1988) focus on political branches rather than administrative agencies, which may be limited in understanding capacities related to both making and executing policy.

From the public administration perspective, scholars point to more practical aspects of policy implementation, such as rulemaking processes and enforcement, to understand administrative capacities. Capacities are then less about available resources or political officials and more about how decision-making is structured both formally (e.g., rules, incentives) and informally (e.g., culture) by organizations during policy implementation (Matland, 1995; McDermott, 2006; Andrews & Boyne, 2010; Brinkerhoff & Morgan, 2010). For instance, Pautz and Rinfret (2013) argue that the regulatory style of the "Lilliputians" (i.e., front-line players, street-level bureaucrats) is essential in determining how effective environmental policies are actually put into practice. Although political science and public administration perspectives differ on theoretical underpinnings of administrative capacities, they are largely in agreement that capacities to manage policies hinge on how choices are made when translating political will into administrative reality. Given that, the task in which we are concerned here is environmental protection. Therefore, our interest is in which capacities contribute to the ability of public agencies to do so, which is a distinct factor from state leaders wanting to protect the environment.

To that end, we can identify three general functions necessary for the operationalization of environmental protection: 1) policymaking; 2) managing information; and 3) creating accountability. First, governments and their subordinate agencies should be able to establish regulations restricting how, when, and where pollutants may be released and the corresponding administrative procedures that allow stakeholders to participate in the design and implementation of regulations (e.g., participatory planning) (Bowman & Kearney, 1988; Egeberg, 1999; Huber & McCarty, 2004; Peters, 2015; Wu, Ramesh, & Howlett, 2015). Second, agencies should be able to collect information on the behaviors of actors who are contributing to environmental conditions, in both positive and negative ways, in order to understand the nature of the problem as it exists within their jurisdictions and reduce any information asymmetries between polluters and regulators (Potoski, 2001; Potoski & Woods, 2001; Whitford, 2002; Brown & Potoski, 2003b). Third, agencies should be able to establish processes and procedures that create accountability both internally and externally in order for both front-line operators and regulatory target populations to make decisions that are consistent with the policies (Walker & Brewer, 2009; DeHart-Davis, 2009; DeHart-Davis, Chen, & Little, 2013). This includes both compliance with guidelines and best practices, as well as enforcing regulations through some means (e.g., coercion, persuasion), so that policy actors comply and do not undermine agency legitimacy (Sappington, 1991; Waterman & Meier, 1998; Miller, 2005).

Building these capacities is not an easy task though and tends to require an investment of both political will and resources. For instance, most tasks associated with environmental protection are both highly technical and asset specific, which creates significant capital costs related to both training a workforce and obtaining specialized equipment. In many cases, these capital costs can be balanced out once an appropriate economy of scale or scope is achieved and efficiencies are realized after reaching a critical mass of operations. However, this disadvantages smaller agencies that may never reach such a tipping point in their operations, as well as larger agencies that can quickly become overburdened by the costs of coordinating large-scale, diverse operations. Additionally, understanding how to effectively use resources tends to be easier when they are already available; that is, someone can read books about driving a car, but true capacity as a driver is built from experience. Consequently, agencies with more experience and resources tend to have more capacity than other agencies, so states with long legacies of environmental protection or established regulatory regimes tend to be better at it than other states. However, this can also make agencies less flexible or adaptive (Callan & Thomas, 2001; Brown & Potoski, 2003a, 2003b; Bikker & van der Linde, 2016)

Whether created actively or passively, organizations may have self-imposed constraints on developing capacities. From one perspective, we may identify these as a function of organizational cultures (i.e., "we do it this way, because that's the way it's always been done!") (March & Olsen, 1989). Within organizations, people are socialized into repeating patterns of behaviors and beliefs about how tasks should be carried out. For instance, if the informal standard operating procedure is to not make an official report of a violation without providing a verbal warning first, it becomes the de facto policy in practice, regardless of the formal written policy. To this end, the norms of policy establish expectations for behaviors both within agencies and among the regulatory target population, which further fortify this approach as the socially acceptable way to enforce policies. Typically, these patterns begin with logical, reasonable choices, but over time they can become antiquated. But if they are ingrained within organizational cultures, the "way we have always done it" may be a more powerful argument than "times have changed." This is particularly important for agencies led by incumbent administrators who have served for decades and may not be enthusiastic about questioning whether processes they helped develop are flawed (Mahler, 1997; Brown & Osborne, 2005; Moynihan & Landuyt, 2009; Peters, 2012).

From another perspective, we may discuss these in terms of uncertainty and risk (i.e., "we don't know what's going to happen, so why risk it?") (Wood & Bohte, 2004). Organizations by their shear nature rely on the formalization of procedures, processes, and operations to function. Formalization institutionalizes practices and behavioral norms that are necessary for stability, especially as people come and go over time. Formalization also creates a degree of certainty in what should happen. But that sense of certainty also makes organizations risk adversities due to the inherent uncertainty that comes along with change, including the potential for unintended consequences. That is, the more certain decision-makers are of what will happen under normal operations, the less likely they are to take actions that create uncertainty (Williamson, 1999; Lubell, Mewhirter, Berardo, & Scholz, 2017). To this end, some argue that legislatures use their rulemaking authority to make restrictive regulations for how agencies operate in order to reduce their own uncertainty in how policies may be implemented in the future. Essentially, legislators want to see policies operationalized in a way that aligns with their interests, so they create constraints on what agencies can do in order to institutionalize certain norms of administrative behavior (Potoski, 1999, 2002; Potoski & Woods, 2001; Wood & Bohte, 2004).

So, why does capacity matter? In the most basic sense, without administrative capacity, it is unlikely that policy actors will be able to protect the environment, even if the political incentives exist. In order to affect change, political will has to be operationalized; otherwise, it is just a set of ideas that gets talked about among policy actors. If those ideas are operationalized in effective ways though, they change how people interact with the environment, and over time it leads to better environmental conditions. In other words, without the capacity to make policies that regulate the emission of pollutants into the environment, there will no standards set for what is or is not acceptable environmental behavior. Without the capacity to enforce compliance with said policies, target populations may not be willing to change their behaviors that are damaging the environment. Without the capacity to monitor environmental conditions or target populations, regulators will not understand how to craft policies in a way that elicits the needed behavioral changes or even who is engaging in negative environmental behaviors. Thus, these capacities are an essential causal link between wanting to protect the environment and taking action to protect the environment.

Policymaking

One of the most essential state administrative capacities is the ability to make policies. Policies come in many different forms, including legislation, executive orders, or the informal norms of practice, but in all forms, policies are a devised solution to a public problem that carries the weight of government (Stone, 2012; Smith & Larimer, 2016). Of course, without these solutions, environmental problems linger and tend to be exacerbated over time, so benign neglect becomes as much of a policy choice as active environmental management

(Ben-Zadok & Gale, 2001; Chiabai et al., 2011; McConnell & T'Hart, 2014). That is, policies represent how elected and/or appointed officials have chosen to solve a pressing environmental problem, or not solve it. Notably, in some cases, officials may choose to ignore a problem because they lack the resources (including time and energy) to develop a politically feasible and technical sound solution to it (Herweg, Zahariadis, & Zohlnhofer, 2018). Within environmental federalism, states are given significant discretion to adapt national standards or guidelines (grounded in federal law or EPA rulemaking processes) to the unique socioeconomic, political, and technical circumstances in their jurisdictions (Crotty, 1987; Scheberle, 2005). By extension, state policymaking capacities fundamentally affect how national environmental programs are operationalized at the state-level, and importantly, how state leaders choose to apply broad national standards to community-level problems.

Scholars rigorously debate what policymaking capacity actually entails, but at its core, it is the capacity of actors to understand, analyze, and apply technical knowledge to policy problems (Janicke, 1997; Wu, Ramesh, & Howlett, 2015). The policy process is complex and multi-faceted, where problem severity and solution feasibility are socially (or politically) constructed by the competing perspectives of different policy actors (Schneider, Ingram, & deLeon, 2014; Herweg, Zahariadis, & Zohlnhofer, 2018). Within this framework,

> we often don't know what the problem is; its definition is vague and shifting. Distinguishing between relevant and irrelevant information is problematic. . . . Choice becomes less an exercise in solving problems and more an attempt to make sense of a partially comprehensive world . . . [and] who pays attention to what and when is critical.
>
> *(Zahariadis, 2014, p. 28)*

Therefore, it is very easy for policymakers to become overwhelmed by the ambiguous and chaotic context in which they operate, which may cause them to make less than optimal policy choices or overlook policy problems altogether. By extension, those states that have the policymaking capacity should have the requisite resources in place to navigate these challenges and develop "good" solutions to environmental problems that meet the unique character of their communities.

So, what does policymaking capacity look like? Because of the rather ambiguous nature of the concept, policymaking capacity is rather difficult to identify (Wu, Ramesh, & Howlett, 2015). While policy capacity functions at the individual, organizational, and system levels, a significant portion of recent studies on policy analysis has focused "specifically on the ability of individuals working in public policy organizations to produce sound analysis to inform policy-making activities" (Howlett, 2015, p. 174). In other words, policymaking capacity tends to come down to the population of professionals working within a policy area, so that states with large professionalized workforces focused on environmental policy issues tend to have more capacity for analyzing and developing policies that match local needs. Thus, policymaking capacity is not just about the production of policies, but rather about the quality of those policies. For instance, Hertel-Fernandez (2014) finds that state legislatures were more likely to enact "model" bills if they have fewer policy resources and by legislators with less policy expertise. In this case, model bills were written by national lobbying organizations hoping to push their policy goals through state legislative processes. In general, model bills lack the nuance to match a policy solution with specific problems within state jurisdictions, so while they are victories for advocacy organizations, they tend to be poor policy choices.

Given this conceptualization, policymaking capacity is fundamentally tied to the ability of state officials to make "good" policy choices for the environment. At the state-level, two factors stand out as being illustrative of comparative differences across the nation: legislative professionalism, and breadth of the environmental policy workforce. In terms of the former, numerous scholars have examined how differences in legislature salaries, resources, and time spent on legislative activities affect their ability to "perform its role in the policymaking process with an expertise, seriousness, and effort comparative to that of other actors in that process" (Mooney, 1994, p. 71). In general, some states rely on citizen legislatures in which legislators serve part-time and with few resources, so they are limited in their ability to understand the policy challenges in front of them. In terms of the latter, states vary widely in the size of the workforce used to manage natural resources, with some states employing enough public servants to reach economies of scale and scope in specialization so that informed decisions can be made within administrative agencies. In turn, workforces expand to include more functions (i.e., pollution prevention, conservation), and the breadth of expertise available for policymaking also expands, which can be used as a resource for elected officials. Although this does not account for the personal talents or abilities of elected or appointed officials in states, it does provide a general idea of the scope of resources available for environmental policymaking across states.

Map 4.1 presents the geographic distribution of our policymaking capacity index.[1] States with higher scores have more professional legislatures and employees working in natural resource agencies. It appears that the policymaking capacity in the Western states largely dwarfs that of other regions, with Alaska and California having the highest capacities in the nation. This is likely a function of the high proportion of natural resource amenities in these states, which would naturally result in states' investing in capacity to manage those assets. States in other regions also display a rather high capacity in this area, including Pennsylvania and New York in the Northeast and North Dakota in

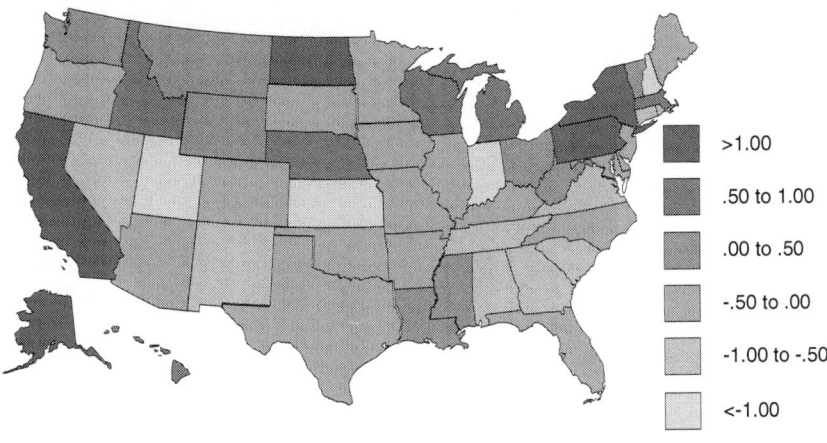

MAP 4.1 Policymaking Index

the Midwest. On the other end of the spectrum, the Southern/Southeastern states have by far the least amount of policymaking capacity, with 13 of 16 states having below average capacity. Interestingly, the states with the least amount overall are not in the South but scattered around other regions, such as New Hampshire in the Northeast, Utah in the West, and Indiana and Kansas in the Midwest. Policymaking capacities in the Northeast and Midwest otherwise appear to be moderate in comparison to the West and South/Southeast.

Managing Information

As a corollary to making "good" policy choices, a second facet of administrative capacity is the ability to effectively manage information. A key obstacle to administrative processes is the limitations of information processing that are experienced both at the individual and organizational level (Lindblom, 1959; Jones, 2003). Herbert Simon (1947) coined the phrase bounded rationality, which refers to rational decision-making as being constrained by the information that is available at the time of the decision. According to Simon, it is impossible to make self-rational decisions without complete information on the circumstances and/or consequences of choices. In other words, lack of information leads to poor decision-making. Certainly, this would suggest that organizations that are capable of collecting and managing information have a higher capacity for making rational decisions based on higher quality and more complete information. Equally as important are information flows in administrative processes, which may lead to information asymmetries (Moe, 1984; Waterman & Meier, 1998). These exist when one party has better information than another party, which in turn allows one party to take advantage of the other through self-interested behavior (e.g., work shirking, payoff maximization). For instance, an unknowledgeable customer may be bilked for high and unnecessary repair costs by an unscrupulous auto mechanic looking to take advantage.

Within environmental policy, state administrative agencies are most clearly in a position to be taken advantage of by polluters, who may exploit an information asymmetry in order to get around complying with regulations. As such, monitoring is an operational necessity within any regulatory system, but especially those in highly technical areas where compliance is not always obvious (Brown & Potoski, 2003b; Miller, 2005). Consequently, states that effectively monitor sources of pollution and manage the corresponding information are also more likely to be able to understand environmental quality issues and enforce compliance with policies. Furthermore, failing to manage information can also put states at a disadvantage in other ways. For example, state policymakers may not be able to negotiate an optimal program adjustment from the EPA if they do not fully understand their own position (Agranoff & McGuire, 2001). Or, appointed officials may not be able to provide the most relevant information to policymakers if information is not collected and managed in a systematic, organized manner. By extension, information management is also an indicator of how well-organized a state's implementation system is, where agencies that struggle to collect and manage the basic data related to their mission are also likely to struggle with more complex aspects of their work (Bretschneider, 1990; Reddick & Frank, 2007).

Conceptually, information management is the ability to collect, organize, and utilize the flow of information to inform decision processes and organizational strategies (Bowman & Kearney, 1988; Wiig, 2002; Christensen & Gazley, 2008; Andrews et al., 2012). So, how is information management capacity operationalized? In order to create this type of capacity, organizations need effective systems, procedures, and rules surrounding what information is to be collected, how it is to be collected, how it is to flow through an organization, and how it is to be utilized by decision-makers. In other words, information management capacity is reflected in whether inspectors consistently inspect facilities, report information to others within the organization, and incorporate past information in making decisions. In the information age, this is largely connected to physical infrastructure in the form of technology (Bretschneider, 1990; Reddick & Frank, 2007; Christensen & Gazley, 2008). However, it is not necessary for agencies to have state-of-the-art gadgets to manage information effectively, but it certainly increases the efficiency of these activities. Given this, one of the most effective ways to understand information management capacities in state environmental agencies is to examine agency performance in terms of collecting and reporting data metrics based on national standards and guidelines.

Although states have significant discretion in designing and implementing programs within their jurisdictions, the EPA provides national guidelines and standards with which states are expected to comply. Starting in the early 2000s, the EPA launched the State Review Framework for Compliance and Enforcement Performance (SRF) to assess and monitor state implementation programs for the CAA, CWA, and RCRA.[2] Specifically, the SRF program is designed "to evaluate state performance to (a) provide a consistent level of environmental and public health protection across states; and (b) develop a consistent mechanism by which EPA Regions, working collaboratively with their states, can ensure that authorized state agencies meet agreed-upon performance levels" (EPA, 2005, p. 1). As such, SRF evaluations provide an adept tool for understanding state implementation practices, as they provide an in-depth assessment of state-level operations within a comparable framework. Currently, SRF evaluations are based on 12 program elements clustered within five constructs: data, inspections, violations, enforcement, and penalties. For our purposes here, the data and inspection elements are particularly telling about inter-state capacities for information management, while the other elements connect closely with the capacity for creating accountability (see the following). An initial review of these reports suggests significant variance in how effectively states collect complete and accurate data, as well as produce complete, high-quality inspections of regulated facilities.

Map 4.2 presents the geographic distribution of our information management index.[3] States with higher scores were rated higher on data management and inspection elements of the EPA's SRF assessment. Similar to the policy-making index, it appears that Western states display a large degree of information management capacity compared to other regions, with Arizona and Nevada having the highest capacities in the nation. On the other hand, it is the Midwest, and not the South, that has fallen behind. More specifically, four of the 12 Midwestern states have below average capacities in this area, compared to three of 13 in the West and six of 16 in the South/Southeast. Overall, the

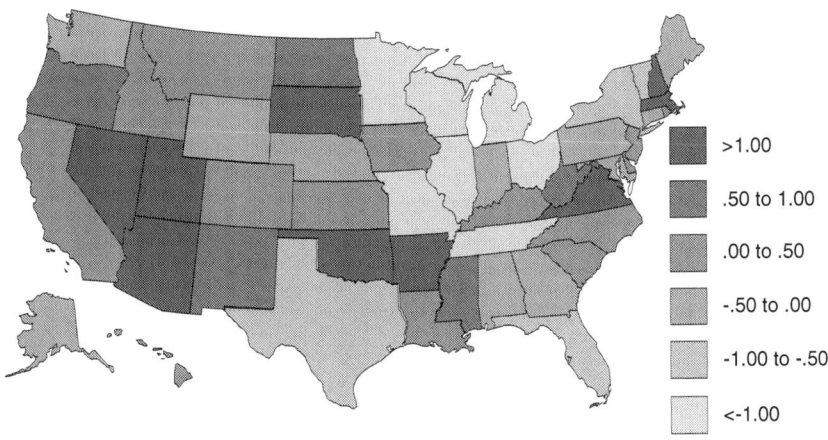

MAP 4.2 Information Management Index

Midwestern states appear to be lagging behind the rest of the nation, with Minnesota, Wisconsin, Illinois, and Michigan having the lowest capacities. On the other hand, the Southern states are particularly interesting, given the relatively low degree of political incentives (see Chapter 4) and the historic low levels of state institutional capacities (Bowman & Kearney, 1988; Lester, 1995). As such, this may be partially explained by Southern states serving as a frequent target of environmental lawsuits and conflicts with the national government over compliance with regulatory standards. In other words, developing capacity in this area may be a coping mechanism to combat challenges to existing environmental institutions in those states. Finally, the Northeastern states are relatively moderate in comparison to the rest of the nation.

Creating Accountability

A final component of administrative capacity is the ability to create accountability with policies. In the most common sense, accountability can be seen as "associated with the process of being called 'to account' to some authority for one's actions" (Mulgan, 2000, p. 555), or "a liability to reveal, to explain, and to justify what one does" (Scott, 2000, p. 40). Both of these definitions imply a social exchange by two or more parties in which one is asserting the rights of authority over another to impose obligations, and the other is responsible for carry out those obligations. Within the modern public service, these obligations tend to connect with serving the public interest, upholding the rule of law, and the institutional checks and balances by which responsible citizens, bureaucrats, and elected officials comply (Mulgan, 2000). Notably, accountability, authority, and responsibility are inherently linked with accountability ultimately being a question of who is responsible for what and who has control for mandating that responsibility and ensuring its adherence to the public interest (Scott, 2000). Within democratic governments, a key element of accountability is the capacity to compel policy actors to modify their behavior in accordance with the public interest in order to focus behavior on achieving desirable societal outcomes (Bovens, Schillemans, & Hart, 2008).

A core concern here is to whom accountability is rendered. In other words, who is responsible for exercising the public interest and who is responsible for taking account of the behavior of policy actors and responding with compelling force, when necessary, in order to alter said behavior (Koliba, Mills, & Zia, 2011). Given the dynamics of environmental federalism, the role states play puts them in the position to "take account" of the behavior-regulated facilities to ensure that they are complying with prescribed behavioral norms (as articulated in regulations) for managing pollution, and, when necessary, to compel those facilities to change their behavior in order to comply with regulations. States also play a role as one who is responsible to their citizens for exercising their own obligations under the law, but also to the national government who delegates authority over federal environmental programs. In other words, states are in a position where they must create accountability internally within agencies and externally with regulated facilities, which complicates their roles (Ruffing, 2015; Fowler, 2019). In both cases, this is a function of states' ability to compel environmental policy actors to conform to policies chosen through democratic processes. In practical terms, these activities include proper monitoring and enforcement of environmental regulations, which is largely synonymous with the practical capacities on which public administration scholars have previously focused. Thus, this dimension is grounded in administrative theory and provides a counterbalance to other dimensions that are grounded more in political science and public policy, respectively.

Most of the basic practices used to create accountability are tied to assumptions from the principal-agent theory, where we assume that agents are likely to drift from their mission, and principals may be just as likely to fail to enforce compliance. For instance, monitoring, reporting, performance evaluation, and incentives or disincentives are all routinely analyzed within the context of creating accountability and controlling agents (Waterman & Meier, 1998; Chinander, 2001; Whitford, 2002; Miller, 2005; Schillemans & Busuioc, 2015). Nevertheless, more state-of-the-art approaches emphasize a shift from prescribing actions to focusing on results as a way to cope with the more complex interorganizational relationships that emerge during the governance era (May, 2007). In either case, if states choose not to hold regulatory facilities accountable for their actions, the legitimacy of environmental regulations is undermined, and those decisions made in the policy process never manifest in the real world. But, if states also fail to hold themselves accountable to the public or the EPA in implementing these policies, they are likely to drift from the mission of environmental protection, and instead choose to serve bureaucratic or political self-interests (Schillemans & Busuioc, 2015). In either case, we should then expect a higher level of environmental protection in practice if states are able to create a high level of accountability both internally and externally.

Given this conceptualization, we can identify three essential elements of creating accountability. First, states should be able to identify behaviors that are out of compliance, which is most readily accomplished through monitoring and inspection programs that identify regulatory violations. Second, states should be able to employ tools in response to noncompliant behavior in order to compel violators to comply with regulations. These tools may not be "one-size-fits-all," and many agencies use a variety of compliance enforcement techniques. Importantly, it is not the use of the tools themselves that creates accountability, but whether they are employed in order to return behavior to a state of compliance. That is, the end goal is regulatory compliance, so whichever tools achieve that are the right ones (e.g., incentives versus disincentives). Third, in order to maintain legitimacy as an objective authority seeking the public interest, states should also be consistent with the application of their responses. Fortunately for our purposes here, the EPA's SRF evaluations also collect data related to monitoring and compliance enforcement that can provide significant insight into these elements. Similar to the results for information management capacities, an initial review suggests significant variance in how effectively states monitor for regulatory violations, response to those violations with enforcement actions, and levy penalties against violators.

Map 4.3 presents the geographic distribution of our creating accountability index.[4] States with higher scores were rated higher on enforcement, violations, and penalty elements on the EPA's SRF assessment. Similar to the policymaking and managing information indices, it appears that Western states display a large degree of capacity for creating accountability within environmental programs compared to other regions. Although states outside the West (e.g., New Hampshire and Virginia) have higher scores, only one Western state (i.e., Arizona) has below average capacity in this area. On the other hand, the Midwest is again lagging behind other regions, with all but two states (i.e., Kansas and South Dakota) having below average capacity; that is in comparison to other regions, like the South/Southeast, where five of 16 states have below average capacity and the Northeast where four of nine have below average capacity. Overall, the Midwestern states seem to struggle to create accountability within their programs, especially states such as Illinois, Michigan, Nebraska, Ohio, and Wisconsin, which have the lowest capacity in the nation. On the other hand, the Southeastern and Northeastern states are relatively moderate in comparison to the rest of the nation, with a few exceptions. For instance, in the South, Oklahoma and Virginia have among the highest capacity, while Kentucky has among the lowest; in the Northeast, New Hampshire has among the highest and Pennsylvania among the lowest.

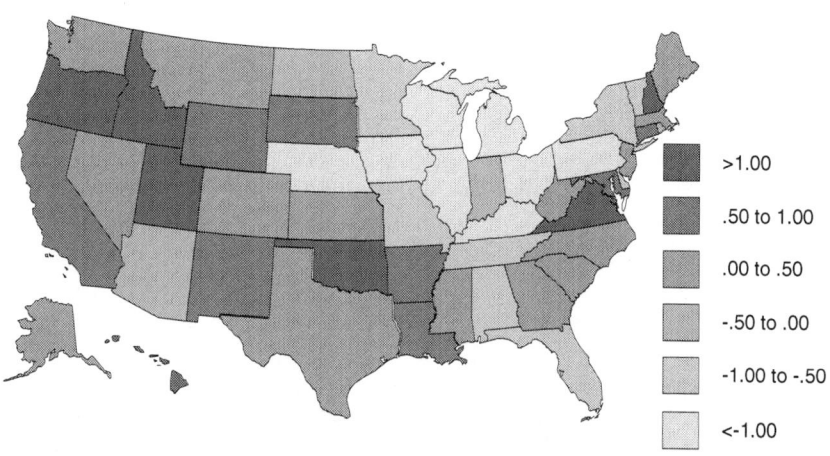

■	>1.00
■	.50 to 1.00
■	.00 to .50
■	-.50 to .00
■	-1.00 to -.50
■	<-1.00

MAP 4.3 Creating Accountability Index

Administrative Capacity Index

As the discussion of policymaking, managing information, and creating account-ability indicate, there is a lot of variation in the administrative capacities by which states provide public services, and those dimensions reveal different limitations for how states protect the environment. Given this, I created the administrative capacity index[5] (see Map 4.4) by combining these measures in order to understand which are most capable of environmental protec-tion. Unsurprisingly, the map shows a concentration of capacities in the West, a dearth of capacity in the Midwest, and moderate levels through the Southeast and Northeast. Of course, there are notable exceptions to those regional gen-eralizations. For instance, in the Midwest, North Dakota and South Dakota display relatively high levels of capacity. Looking at individual states, those with the lowest scores were almost exclusively from the Midwest, with Illinois, Wisconsin, Minnesota, Michigan, Ohio, and Missouri having the lowest in the nation. However, also in this category are Alabama and Tennessee. On the other end of the spectrum, states with the highest capacity ratings include Alaska, California, Idaho, and Hawaii from the West, Oklahoma and Virginia from the South, Massachusetts from the Northeast, and North Dakota from the Midwest.

The administrative capacity index provides us insights into how states dif-fer in their capabilities in protecting the environment; or, alternatively, their ability to operationalize environmental protection. For states at the higher end of this scale, there is an expectation that environmental protection is well-executed, while states at the lower end of the scale likely struggle to implement programs, obtain or organize resources, or make policies that result in positive

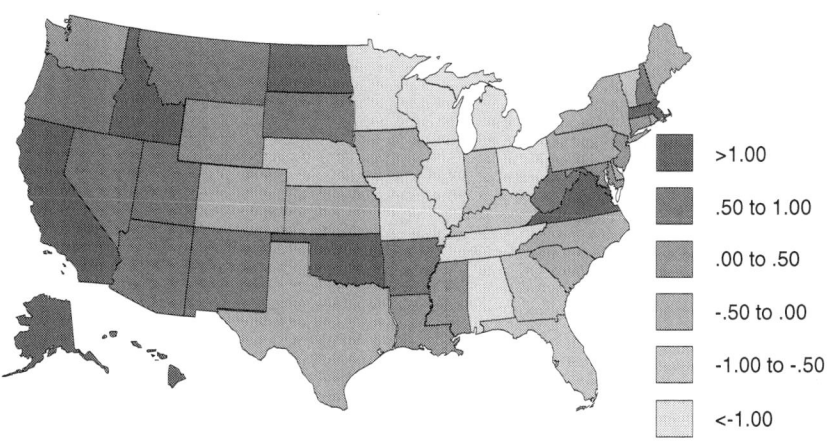

■	>1.00
■	.50 to 1.00
■	.00 to .50
▨	-.50 to .00
▨	-1.00 to -.50
□	<-1.00

MAP 4.4 Administrative Capacities Index

change. These variations in capabilities again highlight both the pros and cons of decentralization as a core tent of environmental federalism. Decentralization is based on the presumption that states are able to provide public services in a way that aligns with community goals and/or needs; however, there are numerous states that may be incapable of doing so. While this may not be a tragedy for states that wish to maintain a low level of environmental protection, there are some states that desire better environmental conditions but lack the tools to achieve it. As such, the administrative capacity index provides a counterbalance to the pro-environmental politics index discussed in the previous chapter and greater insight into how states play their role as linchpins in environmental federalism. Given these two dimensions though, we should now shift focus to the implications for environmental federalism as a result of these variations, which we will take up in the next chapters.

Notes

1. I created a composite measure of legislative professionalism and employees in natural resource agencies. First, I used the Squire Index from 2003 and 2015 to calculate the average degree of legislative professionalism in states for the 2000s. The Squire Index is based on legislator salary and benefits, time spent on legislative service, and staff resources (Squire, 2007, 2017). While there are other measures used, the Squire Index is one of the most frequently cited (Bowen & Greene, 2014). The Squire Index is not available for every year, and these years were the most applicable to our time period of interest, but there is a high degree of consistency between years (r = .91 for 2003 and 2015) (Squire, 2017). Second, using data from the Council of State Governments (2019) and the US Census Bureau (2019), I calculated the average per capita rate of full-time equivalent employees serving in natural resources agencies between 2000 and 2010. Third, I calculated z-scores to create a common scale for both variables and combined into a composite measure. Although there are potentially other indicators, I believe these two get at the core of policymaking capacity in the legislative branch and administrative agencies.
2. Following a first round of evaluations between 2004 and 2007 (based on data from fiscal years (FYs) 2001 to 2004), the EPA revised the SRF evaluation instrument, and a second round occurred between 2009 and 2013 (based on data from FYs 2007 to 2011). Although a third round began in 2015 and a fourth round in 2019, complete results are currently unavailable. EPA and ECOS designed SRF evaluations to be collaborative in nature with the EPA and state agencies developing a mutual understanding of issues, causes, and needed courses of corrective actions. In conducting evaluations, EPA personnel review data from the national data system and a sample of state files related to data, inspections, violations, enforcements, and penalties, which are subdivided into 12 program elements. Generally, evaluations rely on quantitative data metrics, but, when applicable, evaluators also provide qualitative comments and recommendations for improvement.
3. I used data from Round 2 of the EPA's SRF assessment, which covered fiscal years between 2007 and 2011. This dataset provides one of the most in-depth analyses of state implementation systems for the CAA, CWA, and RCRA. Round 2 is the most developed and provides insights into implementation performance during the mid to late 2000s. First, I identified six of 12 data elements reviewed that fell under larger assessment constructs related to data management (data completeness, data accuracy, and timeliness of data entry) and inspections (completion of commitments, inspective

coverage, and quality of inspection or compliance evaluation reports). Second, I collected data on the ordinal ratings for each element across CAA, CWA, and RCRA programs. Elements were originally rated on a four-point ordinal scale based on a series of data metrics; I recoded these into a numerical scale (0 = area for state improvement; 1 = area for state attention; 2 = meets expectations or program requirements; and 3 = good practice). Finally, I calculated the state numerical average across the six elements within three programs and then used z-scores to rescale.

4. I again used data from Round 2 of the EPA's SRF assessment, which covered fiscal years between 2007 and 2011. First, I identified six of 12 data elements reviewed that fell under larger assessment constructs related to enforcement (enforcement actions promote return to compliance, timely and appropriate action), violations (identification of alleged violations, identification of significant noncompliance and high-priority violations), and penalties (penalty calculation method, final penalty assessment collection). Second, I collected data on the ordinal ratings for each element across CAA, CWA, and RCRA programs. Elements were originally rated on a four-point ordinal scale based on a series of data metrics; I recoded these into a numerical scale (0 = area for state improvement; 1 = area for state attention; 2 = meets expectations or program requirements; and 3 = good practice). Finally, I calculated the state numerical average across the six elements within three programs and then used z-scores to rescale.

5. Since all three measures were already measured on a common scale via z-scores, I combined into a composite measure by calculating the average for each state. While managing information and creating accountability indexes are strongly correlated ($r = .65$), there was a moderately weak, negative association with the policymaking index ($r = −.12$ and $−.18$, respectively). Given that I included the policymaking index to capture a different dimension than the other two in order to isolate states that may demonstrate capacities in ways that are not well captured by the SRF evaluations, I am not surprised that it has a weak correlation because the other two indexes capture rather specific program-level elements.

References

Agranoff, R. & M. McGuire. 2001. American Federalism and the Search for Models of Management. *Public Administration Review* 61(6): 671–681.

Andrews, R. & G.A. Boyne. 2010. Capacity, Leadership, and Organizational Performance: Testing the Black Box Model of Public Management. *Public Administration Review* 70(3): 443–454.

Andrews, R., G.A. Boyne, K.J. Meier, L. O'Toole, & R.M. Walker. 2012. Vertical Strategic Alignment and Public Service Performance. *Public Administration* 90(1): 77–98.

Ben-Zadok, E. & D.E. Gale. 2001. Innovation and Reform, Intentional Inaction, and Tactical Breakdown: The Implementation Record of the Florida Concurrency Policy. *Urban Affairs Review* 36(6): 836–871.

Bikker, J. & D. van der Linde. 2016. Scale Economics in Local Public Administration. *Local Government Studies* 42(3): 441–463.

Bovens, M., T. Schillemans, & P.T. Hart. 2008. Does Public Accountability Work? An Assessment Tool. *Public Administration* 86(1): 225–242.

Bowen, D.C. & Z. Greene. 2014. Should We Measure Professionalism with an Index? A Note on Theory and Practice in State Legislative Professionalism Research. *State Policy & Policy Quarterly* 14(3): 277–296.

Bowman, A.O'M. & R.C. Kearney. 1988. Dimensions of State Government Capability. *Western Political Quarterly* 41(2): 341–362.

Bretschneider, S. 1990. Management Information Systems in Public and Private Organizations; An Empirical Test. *Public Administration Review* 50(5): 536–545.

Brinkerhoff, D.W. & P.J. Morgan. 2010. Capacity and Capacity Development: Coping with Complexity. *Public Administration & Development* 30(1): 2–10.

Brown, K. & S.P. Osborne. 2005. *Managing Change and Innovation in Public Service Organizations*. New York: Routledge.

Brown, T.L. & M. Potoski. 2003a. Contract-Management Capacity in Municipal and County Governments. *Public Administration Review* 63(2): 153–164.

Brown, T.L. & M. Potoski. 2003b. Transaction Costs & Institutional Explanations for Government Service Production. *Journal of Public Administration Research & Theory* 13(4): 441–468.

Callan, S.J. & J.M. Thomas. 2001. Economics of Scale and Scope: A Cost Analysis of Municipal Solid Waste Services. *Land Economics* 77(4): 548–560.

Chiabai, A., C.M. Travisi, A. Markandya, H. Ding, & P.A.L.D. Nunes. 2011. Economic Assessment of Forest Ecosystem Services Losses: Cost of Policy Inaction. *Environmental & Resource Economics* 50(3): 405–445.

Chinander, K.R. 2001. Aligning Accountability and Awareness for Environmental Performance in Operations. *Production & Operations Management* 10(3): 276–291.

Christensen, R.K. & B. Gazley. 2008. Capacity for Public Administration: Analysis of Meaning and Measurement. *Public Administration & Development* 28(4): 265–279.

Council of State Governments. 2019. *Book of the States* [online]. Available at http://knowledgecenter.csg.org/kc/category/content-type/content-type/book-states [Retrieved January 1, 2019].

Crotty, P.M. 1987. The New Federalism Game: Primacy Implementation of Environmental Policy. *Publius* 17(2): 53–67.

DeHart-Davis, L. 2009. Green Tape and Public Employee Rule Abidance: Why Organizational Rule Attributes Matter. *Public Administration Review* 69(5): 901–910.

DeHart-Davis, L., J. Chen, & T.D. Little. 2013. Written versus Unwritten Rules: The Role of Rule Formalization in Green Tape. *International Public Management Journal* 16(3): 331–356.

Egeberg, M. 1999. The Impact of Bureaucratic Structure on Policy Making. *Public Administration* 77(1): 155–170.

Fowler, L. 2016. Local Governments: The "Hidden" Partners of Air Quality Management. *State & Local Government Review* 48(3): 175–188.

Fowler, L. 2019. Best Practices for Implementing Federal Environmental Policies: A Principal-Agent Perspective. *Journal of Environmental Planning & Management* [advanced online publication].

Hertel-Fernandez, A. 2014. Who Passes Business's "Model Bills"? Policy Capacity and Corporate Influence in U.S. State Politics. *Perspectives on Politics* 12(3): 582–602.

Herweg, N., N. Zahariadis, & R. Zohlnhofer. 2018. The Multiple Streams Framework: Foundations, Refinements, and Empirical Applications. In *Theories of the Policy Process*, 4th ed., edited by C.M. Weible & P.A. Sabatier (pgs. 17–54). Boulder, CO: Westview.

Howlett, M. 2015. Policy Analytical Capacity: The Supply and Demand for Policy Analysis in Government. *Policy & Society* 34(3–4): 173–182.

Huber, J.D. & N. McCarty. 2004. Bureaucratic Capacity, Delegation, and Political Reform. *American Political Science Review* 98(3): 481–494.

Ingraham, P.W., P.G. Joyce, & A.K. Donahue. 2003. *Government Performance: Why Management Matters*. Baltimore, MD: Johns Hopkins University Press.

Janicke, M. 1997. The Political System's Capacity for Environmental Policy. In *National Environmental Policies: A Comparative Study of Capacity-Building*, edited by M. Janicke, H. Jorgens, & H. Weidner (pgs. 1–24). Berlin: Springer-Verlag.

Jones, B.D. 2003. Bounded Rationality and Political Science: Lessons from Public Administration and Public Policy. *Journal of Public Administration Research & Theory* 13(4): 395–412.

Koliba, C.J., R.M. Mills, & A. Zia. 2011. Accountability in Governance Networks: An Assessment of Public, Private, and Nonprofit Emergency Management Practices Following Hurricane Katrina. *Public Administration Review* 71(2): 210–220.

Lester, J.P. 1995. Federalism and State Environmental Policy. In *Environmental Politics and Policy: Theories and Evidence*, 2nd ed. Durham, NC: Duke University Press.

Lindblom, C.E. 1959. The Science of Muddling Through. *Public Administration Review* 19(2): 79–88.

Lubell, M., J.M. Mewhirter, R. Berardo, & J.T. Scholz. 2017. Transaction Costs and the Perceived Effectiveness of Complex Institutional Systems. *Public Administration Review* 77(5): 668–680.

Mahler, J. 1997. Influences of Organizational Culture on Learning in Public Agencies. *Journal of Public Administration Research & Theory* 7(4): 519–540.

March, J.G. & J.P. Olsen. 1989. *Rediscovering Institutions: The Organizational Basis of Politics*. New York: Free Press.

Matland, R.E. 1995. Synthesizing the Implementation Literature: The Ambiguity-Conflict Model of Policy Implementation. *Journal of Public Administration Research & Theory* 5(2): 145–174.

May, P.J. 2007. Regulatory Regimes and Accountability. *Regulation & Governance* 1(1): 8–26.

McConnell, A. & P. T'Hart. 2014. *Public Policy as Inaction: The Politics of Doing Nothing*. SSRN Working Paper 2500010, University of Sydney.

McDermott, K.A. 2006. Incentives, Capacity, and Implementation: Evidence from Massachusetts Education Reform. *Journal of Public Administration Research & Theory* 16(1): 45–65.

Miller, G.J. 2005. The Political Evolution of Prinicpal-Agent Models. *Annual Review of Political Science* 8: 203–225.

Moe, T. 1984. The New Economics of Organization. *American Journal of Political Science* 28(4): 739–777.

Mooney, C.Z. 1994. Measuring US State Legislative Professionalism: An Evaluation of Five Indices. *State & Local Government Review* 26(2): 70–78.

Moynihan, D.P. & N. Landuyt. 2009. How Do Public Organizations Learn? Bridging Cultural and Structural Perspectives. *Public Administration Review* 69(6): 1097–1105.

Mulgan, R. 2000. "Accountability": An Ever-Expanding Concept? *Public Administration* 78(3): 555–573.

Pautz, M.C. & S.R. Rinfret. 2013. *The Lilliputians of Environmental Regulation: The Perspective of State Regulators*. New York: Routledge.

Peters, B.G. 2012. *Institutional Theory in Political Science: The New Institutionalism*, 3rd ed. New York: Continuum International Publishing Group.

Peters, B.G. 2015. Policy Capacity in Public Administration. *Policy & Society* 34(3–4): 219–228.

Potoski, M. 1999. Managing Uncertainty through Bureaucratic Design: Administrative Procedures and State Air Pollution Control Agencies. *Journal of Public Administration Research & Theory* 9(4): 623–640.

Potoski, M. 2001. Clean Air Federalism: Do States Race to the Bottom? *Public Administration Review* 61(3): 335–342.

Potoski, M. 2002. Designing Bureaucratic Responsiveness: Administrative Procedures and Agency Choice in State Environmental Policy. *State Politics & Policy Quarterly* 2(1): 1–23.

Potoski, M. & N.D. Woods. 2001. Designing State Clean Air Agencies: Administrative Procedures and Bureaucratic Autonomy. *Journal of Public Administration Research & Theory* 11(2): 203–222.

Reddick, C.G. & H.A. Frank. 2007. E-Government and Its Influence on Managerial Effectiveness: A Survey of Florida and Texas City Managers. *Financial Accountability & Management* 23(1): 1–26.

Ruffing, E. 2015. Agencies between Two Worlds: Information Asymmetry in Multilevel Policy-making. *Journal of European Public Policy* 22(8): 1109–1126.

Sappington, D.E.M. 1991. Incentives in Principal-Agent Relationships. *Journal of Economic Perspectives* 5(2): 45–66.

Scheberle, D. 2005. The Evolving Matrix of Environmental Federalism and Intergovernmental Relationships. *Publius* 35(1): 69–86.

Schillemans, T. & M. Busuioc. 2015. Predicting Public Sector Accountability: From Agency Drift to Forum Drift. *Journal of Public Administration Research & Theory* 25(1): 191–215.

Schneider, A.L., H. Ingram, & P. deLeon. 2014. Democratic Policy Design: Social Construction of Target Populations. In *Theories of the Policy Process*, 3rd ed., edited by P.A. Sabatier & C.M. Weible (pgs. 105–150). Boulder, CO: Westview.

Scott, C. 2000. Accountability in the Regulatory State. *Journal of Law & Society* 27(1): 38–60.

Simon, H.A. 1947. *Administrative Behavior: A Study of Decision Making Processes in Administration Organizations*. New York: Free Press.

Smith, K.B. & C.W. Larimer. 2016. *The Public Policy Theory Primer*, 3rd ed. Boulder, CO: Westview.

Squire, P. 2007. Measuring State Legislative Professionalism: The Squire Index Revisited. *State Politics & Policy Quarterly* 7(2): 211–227.

Squire, P. 2017. A Squire Index Update. *State Politics & Policy Quarterly* 17(4): 361–371.

Stone, D. 2012. *Policy Paradox: The Art of Political Decision Making*, 3rd ed. W.W. Norton & Company.

Travis, R., J.C. Morris, & E.D. Morris. 2004. State Implementation of Federal Environmental Policy: Explaining Leveraging in the Clean Water State Revolving Fund. *Policy Studies Journal* 32(3): 461–480.

U.S. Census. 2019. *Statistical Abstract Series* [online]. Available at www.census.gov/library/publications/time-series/statistical_abstracts.html [Retrieved January 1, 2019].

U.S. Environmental Protection Agency. 1993. *William D. Ruckelshaus: Oral History Interview* [online]. Available at https://archive.epa.gov/epa/aboutepa/william-d-ruckelshaus-oral-history-interview.html

U.S. Environmental Protection Agency (EPA). 2005. *Evaluation of the OECA/ECOS State Review Framework in Pilot States* [online] Available at https://www.epa.gov/sites/production/files/2015-09/documents/eval-oeca-ecos-state-review-framework-pilot-projects.pdf

Walker, R.M. & G.A. Brewer. 2009. Can Management Strategy Minimize the Impact of Red Tape on Organizational Performance? *Administration & Society* 41(4): 423–448.

Waterman, R.W. & K.J. Meier. 1998. Principal-Agent Models: An Expansion? *Journal of Public Administration Research & Theory* 8(2): 173–202.

Whitford, A.B. 2002. Decentralization and Political Control of the Bureaucracy. *Journal of Theoretical Politics* 14(2): 167–194.

Wiig, K.M. 2002. Knowledge Management in Public Administration. *Journal of Knowledge Management* 6(3): 224–239.

Williamson, O.E. 1999. Public and Private Bureaucracies: A Transaction Cost Economic Perspective. *Journal of Law, Economics, & Organization* 15(1): 306–342.

Wood, B.D. & J. Bohte. 2004. Political Transaction Costs and the Politics of Administrative Design. *Journal of Politics* 66(1): 176–202.

Wu, X., M. Ramesh, & M. Howlett. 2015. Policy Capacity: A Conceptual Framework for Understanding Policy Competences and Capabilities. *Policy & Society* 34(3–4): 165–171.

Zahariadis, N. 2014. Ambiguity and Multiple Streams. In *Theories of the Policy Process*, 3rd ed., edited by P.A. Sabatier & C.M. Weible (pgs. 25–58). Boulder, CO: Westview.

5

PROGRESSIVES, STRUGGLERS, DELAYERS, AND REGRESSIVES

We started this examination (in Chapter 1) by laying out a base assumption: states want to provide environmental protection at the highest level possible within practical limitations. While that sounds rather rudimentary, it is an exceptionally complex statement for two reasons. First, how state leaders identify the highest level possible is likely to fluctuate based on how committed they are to environmental protection. Second, the practical limitations of environmental protection make it difficult, if not impossible, to achieve the highest level possible, so states are significantly constrained in operationalizing their environmental commitment. From the previous two chapters, which provide detailed discussion on both of these points, we can glean that states fluctuate widely on both the relative levels and structures of political incentives and administrative capacities. The natural extension of those chapters is to now consider how those factors overlap to explain state behaviors related to environmental protection. More specifically, it is likely that if we consider states with both high political incentives and administrative capacities, we can identify discernible patterns of environmental outputs as well as relationships with national and local governments that are unique and different from patterns exhibited by states with either low political incentives or administrative capacities.

A Typology of States

Given the institutional barriers that shape their perspectives, states face different contexts in which they manage environmental problems. In considering the growing role of states in environmental policy during the 1990s, Lester (1995) argued that the efficacy of environmental protection will depend on both the political will of states to commit themselves to environmental protection and the institutional capabilities to effectively implement programs. As discussed in previous chapters, state leaders respond to the political incentives available to them in making environmental policy decisions, and administrative capacities dictate the bounds of what states can do in practice. Although times have changed since the 1990s, other scholars have also pursued this line of reasoning as an important point of inquiry, as it relates to the political and administrative contexts of running environmental programs (O'Leary & Yandle, 2000; Koontz, 2002; Travis, Morris, & Morris, 2004; Konisky & Woods, 2012; Fowler, 2016). Based on these two dimensions, which we operationalize with our political incentives index and administrative capacity index in the previous chapters, we can divide states into four types: 1) progressives; 2) strugglers; 3) delayers; and 4) regressives (see Table 5.1).

First, progressive states are those with high enough political incentives that we should expect state leaders to take environmental protection seriously, and with sufficient administrative capacity to operationalize environmental commitment into policies that create positive change. While political incentives and administrative capacities are both a combination of factors, in progressive states, there is enough substance surrounding both of these dimensions to make environmental protection a priority for policymakers and a goal for administrators. Taken in combination, we should largely expect better environmental policy outputs (i.e., less pollution) than other states that are deficit in either area. Furthermore, we should also expect progressive states to serve as the aggressor in intergovernmental conflicts with the national government, as they push national agencies to be more aggressive. In many ways, the national government serves as a moderating force to balance out the variations between states, but we usually think of it as a force that pulls states forward. However, there are also states on the opposite end of the spectrum that are ahead of the national government and are in a position to challenge national policymakers for leadership on environmental policy. This is the position in which progressive states find themselves. Notably, local governments in progressive states are most likely to play subservient roles to states as compliance managers, as state agencies effectively dominant this policy area.

Second, struggler states are those in which the political incentives for environmental protection are not matched with the administrative capacities to operationalize policy. In these states, leaders may want to be more aggressive

TABLE 5.1 Typology of How States Protect the Environment

		Administrative Capacity	
		High	Low
Political Incentives	High	**Progressives**	**Strugglers**
	Low	**Delayers**	**Regressives**

in environmental protection, but they lack the capability of doing so. Given this, environmental protection may be a frequent priority for policymakers, but administrators are likely to struggle (as the name implies) to operationalize those goals. Thus, we should largely expect environmental programs in these states to fall short of producing the type of outcomes that progressive states do, even though they may outperform states with fewer political incentives. We should also expect struggler states to avoid conflict with the national government and instead embrace cooperation in an effort to make up for their limited capacities. As such, local governments are also likely to function as implementation partners, as states seek access to as much additional administrative capacity as possible. In other words, state leaders are likely to find ways to cooperate with any other governmental unit that can help provide environmental protection.

Third, delayer states are those in which state leaders have few political incentives for environmental protection but have the administrative capacities to comply with national guidelines and standards. In these states, managers of environmental programs are likely to do the bare minimum to meet EPA requirements and avoid conflict. As such, environmental programs are likely to perform better than those in states without sufficient administrative capacities, but not nearly as well as progressive states. Given the lack of political incentives, a key driving force for how these states approach environmental protection is in conflict with the national government, or more importantly, avoids conflict with the national government. Instead, they favor the path of least resistance, which is compliance. Consequently, we should expect these states to largely avoid conflict or cooperation with the national government and keep a relatively low profile in environmental federalism. In turn, this creates a unique challenge where states have capacities to implement programs but face political pushback from doing so. In order to overcome this, states engage stakeholders at the community-level and allow them to develop feasible political solutions to environmental challenges.

Finally, regressive states are those in which neither the political incentives for environmental protection nor the administrative capacity to protect the environment exist at sufficient levels. In these states, leaders are largely ambivalent (and possibly antagonistic) towards environmental concerns, and administrative agencies lack the fundamental capabilities necessary to effectively and efficiently manage programs. Therefore, we should largely expect these states to fall far behind the rest of the nation in environmental policy outputs. Furthermore, we should also expect these states to be the thorn in the side of the national government as they consistently fail to comply with national standards, leading to conflict and negative interactions. For these states, compliance with federal regulations is neither a primary political concern nor practically achievable, so they are likely to develop uncooperative relationships with the EPA. Local governments in regressive states are in a somewhat difficult position,

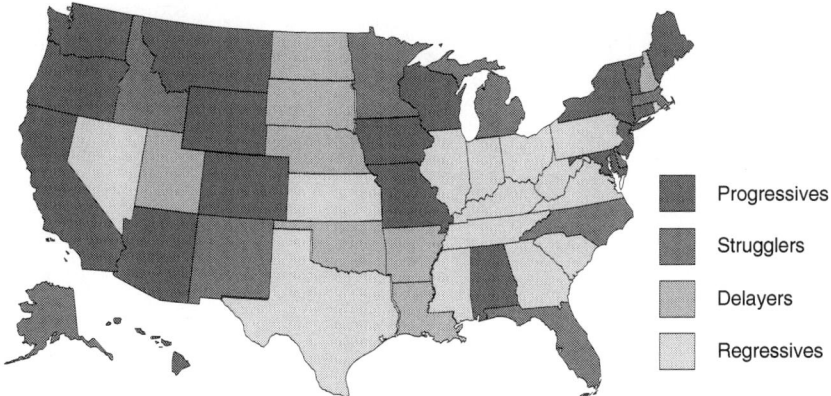

Progressives

Stragglers

Delayers

Regressives

MAP 5.1 Map of States by Type

as state agencies fail to address environmental needs and local communities are largely ambivalent. As such, in some cases, local governments are likely to struggle with complying with national standards, as the absence of state supports puts them in a vulnerable position.

To understand which states fall into each category, we combined our pro-environmental politics and administrative capacity indices (see Map 5.1).[1] Our classification resulted in a fairly balanced distribution across categories, with 15 progressive states, 12 struggler states, 11 delayer states, and 12 regressive states. Regional trends are not particularly surprising given the discussion of political incentives and administrative capacities in previous chapters. The majority of progressive states (11 of 15) are located in the West, while the remaining are in the South/Southeast (Maryland) or Northeast (Connecticut, Massachusetts, New Jersey). Additionally, only two Western states (Nevada and Utah) are not classified as progressive states, so there is a notable regional concentration for this category of states. On the other hand, there is far more regional diversity across the other categories. For instance, there are both Southern/Southeastern and Northeastern states in every category. In contrast, there are no progressive Midwestern states, and the majority are either stragglers or regressives. Breaking out of the somewhat myopic census regions and looking more broadly at the map, there is certainly a concentration of regressive states in the Ohio River Valley states (i.e., Pennsylvania to Tennessee). Otherwise, there are few notable regional trends that extend beyond two or three states.

Figure 5.1 graphs state scores on both indices. Notably, there are many states that are "on the margins" in our classification. Our indices used to measure these concepts are calculated as z-scores, so that zero represents the mean value, +1 represents one standard deviation above the mean, and −1, one standard deviation below. By nature of using z-scores and assuming a standard distribution, there is a tendency for states to group around the mean (or center) of

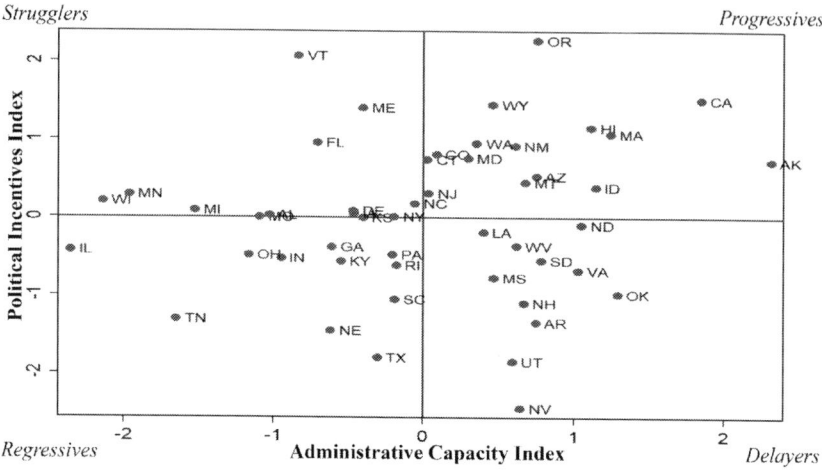

...

FIGURE 5.1 Scatterplot of Political Incentives and Administrative Capacity Indices

the graph. This, of course, complicates our attempt to draw clear distinctions around certain types of states. Accordingly, states with a score +/−0.25 standard deviations from the mean are within approximately 10% of the national average, which make them somewhat vulnerable to misclassification. In fact, only 33 states scored above +0.25 or below −0.25 standard deviations on both indices. On the other hand, two states (New York and North Carolina) are within +/−0.25 standard deviations of the mean on both indices. Consequently, it should be noted that this typology is not necessary static, and states have the potential to migrate from one category to another over time (and depending on measurement validity and reliability). We believe that this is an accurate accounting of states for the 2000s, but it is worth noting that some states are "on the margins."

Given the discussion in previous chapters about the complexity of both political incentives and administrative capacities, it is worth taking a minute to examine whether there are specific factors connected to particular categories of states. In other words, is public opinion a more important driver of political incentives in progressive states than advocacy groups? Or, is policymaking a more important driver of administrative capacities in regressive states than managing information? In order to answer these questions, let us first look at Table 5.2, where we use the probabilities that a state would fall into each category as a dependent variable and the six constituent parts (public opinion, advocacy groups, comparative policy, policymaking, information management, and creating accountability) of our indices as predictor variables.[2] While findings show that public opinion and advocacy groups, on one hand, and creating accountability, on the other, are the most important predictors, findings also indicate that different elements are more important in different types of states. For instance, it appears that the influence of public opinion and advocacy

TABLE 5.2 Regression of Probabilities of States Falling Into Each Category[3]

	Progressives***	Strugglers***	Delayers***	Regressives***
Public Opinion	.16**	.11	−.12*	−.15*
Advocacy Groups	.18**	.09	−.16*	−.11
Competition	−.005	−.14*	.008	.14
Information Management	.01	−.09	.13*	−.05
Creating Accountability	.21**	−.16*	.12	−.17*
Policymaking	.16*	−.12	.03	−.07
Constant	.31	.24	.23	.23
R-squared	.61	.43	.52	.44
Adjusted R-squared	.56	.35	.45	.36
N	48	48	48	48

Note: Ordinary Least Squares (OLS) regression is used, with statistical significance levels indicated by *<.05, **<.01, and ***<.001.

groups is relatively equal in predicting the probability that a state falls into the progressive or delayer categories. Given that coefficients are positive for progressives and negative for delayers though, it would appear that the political incentives in states that fall into these respective categories are structured similarly but are inversed from each other.

On the other hand, only public opinion is a statistically significant predictor for regressive states, and comparative policy for delayer states. In general, this would suggest that political incentives in struggler or regressive states are structured differently than those in progressive or delayer states. Looking at administrative capacities, it appears that the probability of creating accountability is relatively equal in predicting whether a state falls into the struggler or regressive type. On the other hand, creating accountability in combination with policymaking capacity determines whether a state falls into the progressive type, while managing information is the determining factor for delayers. Most interestingly, it appears that the key difference between strugglers and regressives is whether political incentives are driven by comparative policy or public opinion, and the key difference between progressives and delayers is whether administrative capacities are driven by policymaking and creating accountability or managing information. From this, we can then conclude that it is not simply the degree of political incentives or administrative capacities that dictates how states operate, but rather the structure of those incentives and capacities. In other words, political incentives and administrative capacities in progressive states are structured in a way that is most advantageous to environmental protection, while in regressive states, they are structured in a way that is least advantageous.

Of course, these findings beg the questions of which factors are most important for individual states. Maps 5.2 and 5.3 show the factor that is the most accurate political incentive and administrative capacity predictor, respectively, for

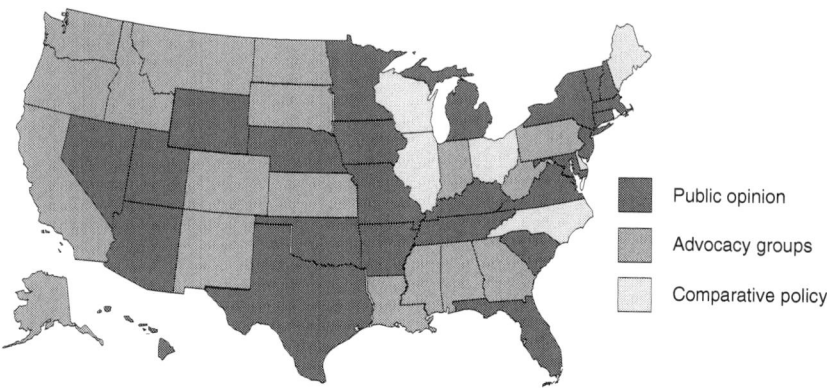

MAP 5.2 Most Accurate Predictors of Type by State, Political Incentives

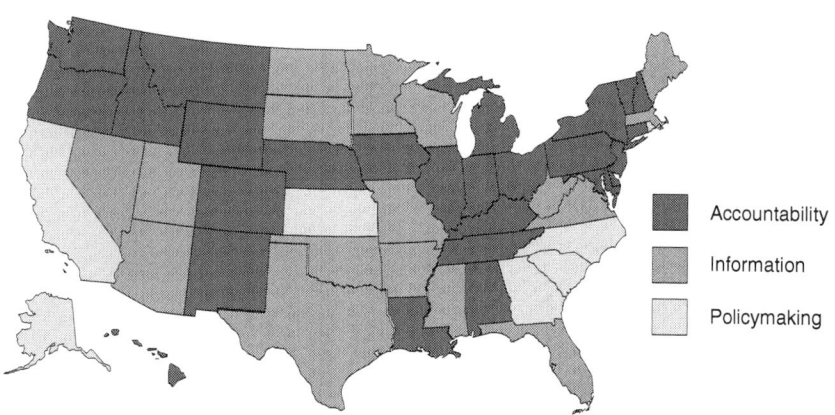

MAP 5.3 Most Accurate Predictors of Type by State, Administrative Capacities

the actual category into which each state fits.[4] On the political incentives side, public opinion is the most accurate predictor for 26 states and advocacy groups for 18 states, while competition is the most accurate for six states. Given the role of public opinion in political decision-making in a democracy, it should not be surprising that it is closely tied to political incentives in the majority of states. Interestingly, the map hints at a few regional patterns where advocacy groups supplant public opinion as the primary driver of political incentives. Most of these are Pacific Ocean or Rocky Mountain states, but there also appears to be a pocket along the Gulf of Mexico, which may be more of a result of weak public opinion than strong advocacy groups. On the administrative capacity side, creating accountability is the most accurate for 26 states and managing information for 17 states, while policymaking is the most accurate for seven states. Few regional patterns seem to appear though, which is fairly consistent with the lack of pronounced regional trends for administrative capacities in general.

As we look at the intersection between variables, we find a few interesting trends. For instance, Connecticut, Hawaii, Maryland, and New Jersey appear to be direct counterparts to Kentucky, Nebraska, and Tennessee in that all are driven by public opinion and accountability, but the former group is above average on both components and are progressive states, and the latter group is below average on both components and are regressive states. Additionally, it appears that there is a some regional clustering. For example, Colorado, Idaho, Montana, New Mexico, Oregon, and Washington are all progerssive states that are driven by advocacy groups and creating accountability; while Connecticut, Maryland, and New Jersey are progressives driven by public opinion and creating accountability. Otherwise, there is a large degree of variation across regions. In general, these maps further hint at how political incentives and administrative capacities are structured differently in states, which is an important factor in considering the obstacles to environmental protection. As findings from Table 5.2 indicate, there are some distinct differences across state categories, which may help us understand how and why there are variations in environmental protection across states. Given this, we should now consider whether this typology is useful in explaining outputs from environmental protection so we can further validate our inferences here.

Explaining Patterns of Pollution

Let us now consider the first part of our expectations outlined earlier: do political incentives and administrative capacity explain environmental protection outputs? In order to answer this question, we should look at this from a variety of perspectives. To start, nearly 36% of toxic chemical releases during the 2000s occurred in the 12 regressive states, as compared to only about 23% in the 15 progressive states. Furthermore, on average, regressive states accounted for almost 3% of pollution individually, compared to about 1.5% for progressive states, 1.7% for struggler states, and 1.9% for delayer states. However, given that environmental impacts of pollution tend to be a function of concentration, this is somewhat limited in telling us about the quality of environmental protection. Turning to Table 5.3, we see the mean pollution concentration

TABLE 5.3 Mean Pollution Concentration by State Category

	*Pollution Concentration**	*Air Pollution Concentration**	*Water Pollution Concentration**	*Land Pollution Concentration**
Progressives	1475.23	366.25	80.37	70.38
Strugglers	2375.26	774.42	121.38	55.67
Delayers	2410.16	582.38	105.03	115.72
Regressives	3950.70	1154.60	177.36	141.11

Note: A one-way ANOVA is used to test for statistical significance of distributions, while statistical significance levels are indicated by *<.001.

TABLE 5.4 Correlations Between Index Variables and Pollution Concentration

	Pollution Concentration	Air	Water	Land
Political Incentives Index	−.29**	−.14**	−.11**	−.17**
– Public Opinion	−.06*	−.02	.02	−.05
– Advocacy Groups	−.29**	−.23**	−.22**	−.02
– Competition	−.01	−.01	.04	.08*
Administrative Capacity Index	−.27**	−.23**	−.12**	−.08*
– Policymaking	−.07*	−.13**	−.12**	.00
– Information Management	−.24**	−.11**	−.03	−.02
– Creating Accountability	−.17**	−.21**	−.10**	−.12**

Note: Pearson's r is used to test for correlations, and a one-tailed test for statistical significance. Statistical significance levels are indicated by *<.01 and **<.001.

by state category.[5] Across the board, there is a clear pattern where progressive states have the lowest mean pollution, and regressive states have the highest. To put these numbers in perspective, progressive states have between 50% and 63% lower pollution concentration than regressive states, while struggler and delayer states have between 40% and 60%, or 18% and 39%, respectively. In other words, there are nontrivial differences between how much pollution is concentrated in different categories of states.

Now let us turn our attention to whether our indices are associated with pollution concentrations. Table 5.4 shows correlations between our indices, and their constituent components, and state-level pollution concentrations. With a few exceptions for water and land pollution, there is a definitive negative correlation between pollution concentrations and our index variables, which on its face would indicate that these factors are connected to environmental conditions in that increases in the index variables are associated with decreased pollution. Notably, our indices have stronger correlations than the constituent components, suggesting that the combination of factors is a stronger indicator than any single factor. This, of course, is part of the motivation for constructing the indices. While several constituent components display weak positive or negative associations, both indices have a strong relationship with general pollution concentration, although those relationships are weaker with the individual types of pollution. So far, we seem to be able to conclude that political incentives and administrative capacities do indeed explain environmental protection outputs, at least in the form of state-level pollution concentration. But, these have been relatively unsophisticated ways to answer our question, so now let us move on to some more sophisticated methods.

Table 5.5 presents results from four regression models. In Model 5.1, we see that there is a negative, statistically significant (i.e., generalizable) relationship for our political incentives and administrative capacity indices, which means increases in either index is associated with decreases in pollution concentrations.

TABLE 5.5 Regression of Pollution Concentrations[6]

	Model 5.1	Model 5.2	Model 5.3	Model 5.4
Political Incentives Index	−666.84***		−714.92***	
Administrative Capacity Index	−605.94***		−368.42*	
Progressives		−2475.47*		−1861.81***
Strugglers		−1575.44*		−1107.77*
Delayers		−1540.54*		−1304.68**
Industry			249.41*	244.97*
State and Local Spending			−149.40	−202.26*
Population			.26	.16
Wealth			−58.19	1.40
Constant	2491.03	3950.70	2806.19	4226.26
R-squared	.14	.13	.34	.33
Adjusted R-squared	.13	.12	.33	.32
Durbin-Watson (trans.)	−	−	.83	.81
N-size	550	550	450	450

Note: Statistical significance levels are indicated by *<.05, **<.01, and ***<.001.

Interestingly, given that both indices are measured on the same scale, the higher coefficient for the political incentives index indicates that political incentives have a stronger association with pollution concentration than administrative capacity. In Model 5.2, negative coefficients indicate that progressive, struggler, and delayer states have lower pollution concentrations than regressive states (the comparison category). Notably, it appears that there is little substantive difference between effects for struggler and delayer states, and further examination indicates the comparison is not statistically significant. While both political incentives and administrative capacity are individually important, it appears that the states in these middle categories may not produce substantive differences in environmental protection outputs. Additionally, the R-squared and adjusted R-squared indicate that both models are moderately strong in predicting pollution concentrations.

Nevertheless, these first two models are still relatively unsophisticated, so Models 5.3 and 5.4 include control variables for industry, state and local spending, population, and wealth.[7] Additionally, since this is panel data (i.e., cross-sectional, time-series) and we suspect serial correlation to exist with our socioeconomic variables, we use the Prais-Winsten estimation to correct for this (Beck & Katz, 1995). In general, this model is a more sophisticated way of controlling for all the "white noise" in the data that may obscure our understanding of how political incentives and administrative capacities are associated with pollution concentrations. In Model 5.3, we again see negative, statistically significant coefficients for both our political incentives and administrative capacity indices. In contrast to Model 5.1, the substantive difference between the two has increased significantly and indicates that the same numeric increase in political incentives has roughly double the substantive impact on pollution concentration as it would for administrative capacities. In other words, once we control for socioeconomics and eliminate some statistical white noise from the model, political incentives appear to be more important in predicting environmental protection outputs than administrative capacities, although both are substantive predictors.

In Model 5.4, we also see that progressive, struggler, and delayer states have lower pollution concentrations than regressive states. Coefficients indicate that pollution concentrations in progressive states are approximately 44% lower than in regressive states, while those in struggler and delayer states are approximately 31% and 26% lower, respectively. Additionally, given the findings from Model 5.3, it is not surprising that the separation between struggler and delayer states has become more pronounced, with pollution concentrations in delayer states being approximately 6% lower than in struggler states, compared to an approximately 1.5% difference in Model 5.2. However, further examination indicates that it is still not a statistically significant difference. Taken as a whole, these findings largely support our expectations that there are substantive differences between state categories in environmental protection outputs, which are manifestations of the relative political incentives and administrative capacities. It is also noteworthy that Models 5.3 and 5.4 are relatively strong predictors of pollution concentrations and outperform Models 5.1 and 5.2. In general, this suggests that political incentives and administrative capacities do not impact environmental conditions independently of socioeconomic conditions, which is consistent with arguments from other scholars.

Are Political Incentives and Administrative Capacity Independent?

A key criticism that may be launched against this typology, and by extension these findings, is whether our two dimensions are in fact independent of each other. That is, some may argue that administrative capacity is a function of political incentives and therefore these dimensions are interdependent. From this perspective, one mechanism by which elected officials may operationalize their desire to invest in environmental protection is through providing resources to administrative agencies to grow their capacities. If this is in fact the case, it would limit our ability to make inferences both theoretically about states and empirically from this data by violating important statistical assumptions in some of the models presented. However, we believe that while intertwined, political incentives and administrative capacities are in fact distinct concepts. Administrative capacities are not strictly a function of political incentives, but rather of the internal mechanisms of administrative agencies where talented managers (or the lack thereof) shape operations more than elected officials wielding legislative or executive authorities. Therefore, administrative capacities can exist at a high level in states with few political incentives for environmental protections, or be absent in states with significant incentives. In either case, the causal association between these two dimensions is likely not strong enough to bias our inferences.

But, let's look at the data. Table 5.6 shows the correlation between the political incentives index and its constituent components, and the administrative capacity index and its constituent components. Here, a statistically significant, positive correlation would suggest as political incentives increase, so do administrative capacities. First, the political incentives index only displays such an association with policymaking and actually has a negative correlation with managing information. On the other hand, only advocacy groups have a statistically significant, positive correlation with the administrative capacity index, while public opinion and competition have negative correlations. Finally, public opinion and policymaking are the only two constituent components of both indices that have a statistically significant, positive correlation, and the majority of the other correlations are negative. In other words, there is no evidence that political incentives are positively correlated with administrative capacity. Table 5.7 takes this an additional step by regressing the administrative capacity

TABLE 5.6 Correlations Between Indices

	Administrative Capacity	Policymaking	Information Management	Creating Accountability
Political Incentives	.21	.37**	−.18	.05
Public Opinion	−.01	.28**	−.23	−.16
Advocacy Groups	.26*	.08	.005	.21
Competition	−.12	−.07	−.07	−.01

Note: Pearson's r is used to test for correlations, and a one-tailed test for statistical significance. Statistical significance levels are indicated by *<.05 and **<.01.

TABLE 5.7 Regression of Administrative Capacities

	Administrative Capacity		Policymaking*		Information Management		Creating Accountability	
Public Opinion	−.09	−	.23	−	−.26	−	−.25	
Advocacy Groups	.38*	−	.01	−	.11	−	.35	.07
Competition	−.23	−	−.10	−	−.10	−	−.05	
Political Incentives Index	−	.22	−	.36**	−	−.17	−	
Constant	.001	.00	−.10	.00	−.02	.00	−.03	.00
R-squared	.10	.05	.09	.13	.06	.03	.10	.01
N-size	48	50	48	50	48	50	48	50

Note: OLS regression is used, and statistical significance levels are indicated by *<.05 and **<.01.

index and its constituent components using the political incentives index and its constituent components as predictors. Here, the only statistically significant predictors are advocacy groups for administrative capacity and the political incentives index for policymaking. This would again suggest no systematic evidence that there is a causal relationship between our two dimensions.

If we are going to make inferences about the relationships between our indices though, we must also consider whether they are inserting bias into our statistical models presented here. If this were occurring, it would likely be along one of two avenues: multicollinearity or endogeneity. In general, regression assumes that predictor variables provide additive value in models, rather than significantly overlap in the variance they explain in the dependent variable. If there is a significant overlap between variables, it can bias standard errors and lead to false inferences about the generalizability of results (Chatterjee & Hadi, 2006). We test for this using the Variance Inflation Factor (VIF); the rule of thumb is that VIF scores should be below ten (Fox, 1992). Table 5.8 displays these for our models using both the category probability (Table 5.2) and pollution concentration (Table 5.5) as dependent variables. In both cases, we see that VIF scores are well below reported ranges, so our models are not in violation of assumptions surrounding multicollinearity.

On the other hand, endogeneity occurs when an explanatory variable is correlated with the error term, which again may bias our standard errors. In this case, it may occur if there is a reciprocal correlation between politics, capacity, and pollution concentrations, where politics affects capacity, capacity affects pollution, and pollution affects politics. We test for this using the Durbin-Wu-Hausman test, and results show that endogeneity is not above critical values for Models 5.1–5.4 (Table 5.5) (Nakamura & Nakamura, 1998). In sum, there is no statistical evidence to suggest that our political incentives and administrative capacity indices are associated to any significant degree. By extension, we can then be confident in our ability to make inferences about the relationships between these two dimensions and how states manage environmental programs.

TABLE 5.8 VIF Scores

	Category Probability (Table 5.2)		Pollution Concentration (Table 5.5)	
Political Incentives		1.02	1.02	
Public Opinion	1.23			1.23
Advocacy Groups	1.23			1.23
Competition	1.05			1.05
Administrative Capacity		1.02	1.02	
Policymaking	1.18			1.18
Information Management	1.87			1.87
Creating Accountability	2.06			2.06

A Brief Thought Experiment

At this point, we think it is a fruitful exercise to conduct a brief thought experiment with our data in order to show that these observations have strategic value for improving environmental protection, and not just descriptive value. More specifically, we can largely assume that progressive states are doing a satisfactory job of providing environmental protection and will largely continue to do so regardless of any structural changes to national policy. On the other hand, we can also assume that regressive states are doing a poor job of protecting the environment and will continue to do so regardless of any national policy changes. But what about the 12 delayer and 11 struggler states that produced about 41.5% of toxic releases during the 2000s? Do these findings give us some idea of the potential of these states to do better? Do these findings give us any indication of how the structure of political incentives or administrative capacities may hold states back? The short answer to the last two questions is yes. Note that while this data concentrates on the 2000s, we will move our discussion to the future and respond to climate change in latter chapters, which this thought experiment hopes to inform.

Let us start with West Virginia as an example. While administrative capacities are sufficient, West Virginia is a delayer state that exhibits below average political incentives based on public opinion and advocacy groups but compares well to its neighbors based on the comparative policy dimension. Given that advocacy groups are the most accurate predictor for West Virginia's political incentives, we know that changes there are the most likely to elicit the necessary alterations in the political environment to push state leaders to be more proactive in environmental protection. Specifically, if campaign contributions from environmental groups were to increase to national mean levels, West Virginia would for all intents and purposes be a progressive state, rather than a delayer state. We assume that increases in contributions would not just be a superficial change in our data, but rather correspond with an increase in the political incentives for state leaders to make pro-environmental choices. Based on our data, contributions would have to be increased by more than 1000% in order for that to happen. While that sounds like a dramatic and unpractical rise in contributions, it would only mean that environmental groups would have to donate as much in West Virginia as they do in Iowa or Maryland. Since our findings indicate that progressive states tend to enjoy 19.1% less pollution concentration than delayer states, we would expect that significant environmental improvements would manifest over time from this uptick in advocacy group activities.

Now, let us think about Iowa. Iowa is a struggler state that has sufficient political incentives for environmental protection but not administrative capacities. Although Iowa is above average in managing information, it is below average for policymaking and creating accountability, and the latter is the most accurate for predicting where Iowa falls in our typology. Digging deeper into our data and the EPA's program evaluation, we can determine that the real cause of this lack of accountability is tied to the identification of violations and timely and appropriate enforcement actions. Iowa's Department of Natural Resources struggles in these areas for both the CAA and CWA, suggesting this is an endemic problem and not simply a miscue on program management. Specifically, the EPA has suggested creating enforcement protocols, updating policies, and training employees on best practices as measures to improve enforcement and violation activities. We assume that taking such steps would increase accountability both internally within Iowa's environmental agencies and with regulated facilities to the point that administrators would become more effective in implementing policies, and pollution would decrease. While these steps sound minor, our findings indicate that progressive states tend to enjoy 24.2% less pollution concentration than struggler states, so again we would expect significant environmental improvement to manifest over time.

Are more campaign contributions and best practices all it would take to improve environmental protection in the US? No! The political and administrative contexts of environmental protection are much more complex than this thought experiment would suggest, so neither campaign contributions nor administrative best practices are magic bullets for West Virginia or Iowa or any other state. Real policy and administrative change or environmental improvement does not occur overnight. It requires a significant investment of time, energy, and resources. But what this thought experiment does indicate is that there are certain components that are missing from how political incentives or administrative capacities are structured in certain states, and if those components can be repaired, improved, or replaced; then we can expect better environmental outcomes over time. While this thought experiment is rather simplistic, it does help us to understand why examining obstacles to effectively managing the environment through the federal system is important, and how understanding these obstacles can facilitate better strategic choices.

With that said, states function as part of a system, and the success of intergovernmental policy implementation and environmental management is not wholly rested on their shoulders. Rather, their relationships with other governmental units (i.e., national, local) are extremely important in dictating how these programs operate in the real-world. When programs are marred by conflict, resources are sucked up by in-fighting, uncertainty leads to decision gridlock, and program management becomes exponentially harder. Thus, it is not just the internal elements of states that account for differences in environmental outcomes; it is also their external relationships with the national government above and local governments below. If these relationships are healthy and positive, we should expect programs to run better and for environmental agencies at different federal levels to find ways to work together to solve collective problems, but if these relationships are negative and filled with conflict, we should expect this to serve as a critical obstacle to environmental protection. To this end, we should now turn our lens of inquiry to how states interact with other governmental units at the national and local-levels that are also protecting the environment.

Notes

1. To determine categories for states, I coded states with positive indices as high (i.e., above average based on z-scores) and those with negative indices as low (i.e., below average based on z-scores).
2. I calculated the dependent variable in two steps. First, I ran a multinomial logit model with the four nominal categories as a dependent variable, and the pro-environmental politics and administrative capacities indices as the predictor variables. Second, I predicted the probability that each state would fall into each type, which created four distinct probability estimates.
3. Diagnostic tests indicate no assumptions are violated (Chatterjee & Hadi, 2006). There are no observations for Alaska and Hawaii on the comparative policy variable, so they were dropped from models.
4. I made these determinations using a four-step process. First, using the same dependent variables from Table 5.2, I ran OLS bivariate regression models with the individual constituent parts of the indices serving as a single predictor for each of the four probability outcome variables (i.e., progressive, and so on). Second, I calculated residual errors from each of those models, with a smaller residual error indicating a more accurate prediction of probability. Third, in order to identify how well each index part predicted the correct type, I matched residual errors to states based on what outcome variable they were estimated from and the type each state was classified as (i.e., if a state was classified as progressive, we focused on the residual errors for the model with progressive probabilities. Finally, I compared residual errors across variables and determined the lowest absolute value).
5. I measure pollution concentrations as the total pounds of toxic releases per square mile, and further break this down into air, water, and land based on available reporting categories. I obtained data from the EPA's Toxics Release Inventory (TRI) (EPA, 2019). The TRI program requires companies classified as mining, manufacturing, utilities, waste management, or wholesale trade and that produce more than 25,000 pounds of toxic chemicals a year to report those chemical releases to the EPA, which then compiles those reports into a national database. Primarily, these toxic releases involve pollutants directly regulated by the CAA, CWA, or RCRA and that pose a threat to human health and the environment. However, TRI data does not include pollutant discharges from nonpoint sources (EPA, 2019). This is a relatively common mechanism in which previous scholars have measured environmental protection efforts, environmental policy efforts, and/or environmental conditions at the state-level. In general, these scholars assume that if environmental quality is dependent on toxic chemical releases and environmental policies regulate those releases to control environmental quality, toxic chemical releases are a measure of the impact of environmental policies (Bacot & Dawes, 1997; Bowen & Wells, 2002; Sapat, 2004; Konisky & Woods, 2012; Fowler, 2019).
6. In Models 5.1 and 5.2, we use OLS regression. Diagnostic tests indicate no assumptions are violated, except for a potential issue with serial correlation (Chatterjee & Hadi, 2006). In Models 5.3 and 5.4, I correct for the serial correlation issue by using the Prais-Winsten estimation and report the transformed Durbin-Watson statistic, which falls into acceptable critical values for both models (Beck & Katz, 1995). Note that in Models 5.3 and 5.4, missing expenditure data for state and local governments reduces the n-size to 450.

7. Previous scholarship indicates that socioeconomic factors related to the composition of the economy, economic development, and size of subnational government have substantive impacts on environmental conditions that should be used as control factors when examining environmental policies (e.g., Lester, 1995; Bacot & Dawes, 1997; Sapat, 2004; Konisky & Woods, 2012; Fowler, 2016). Using data from the US Census Bureau (2019) and Bureau of Economic Analysis (2019), I measure: 1) industry as per capita economic production from the manufacturing, mining, utilities, and waste management sectors in 2009 dollars; 2) state and local total per capita expenditures in 2009 dollars; 3) total state population in thousands of people; and, 4) per capita personal income in thousands of 2009 dollars. While there are other socioeconomic, political, and technical variables that other scholars also use as control variables, many of these are either incorporated into our political incentives or administrative capacity indices or measure related concepts, which creates a possibility of unrealized multicollinearity in the model. As such, I reduce my model to the basic socioeconomic factors and assume that my indices and categorization incorporate most of these other factors in one way or another.

References

Bacot, A.H. & R.A. Dawes. 1997. State Expenditures and Policy Outcomes in Environmental Program Management. *Policy Studies Journal* 25(3): 355–370.

Beck, N. & J.N. Katz. 1995. What to Do (and Not to Do) with Time-Series Cross-Section Data. *American Political Science Review* 89(3): 634–647.

Bowen, W.M. & M.V. Wells. 2002. The Politics and Reality of Environmental Justice: A History and Considerations for Public Administrators and Policy Makers. *Public Administration Review* 62(6): 688–698.

Chatterjee, S. & A.S. Hadi. 2006. *Regression Analysis by Example.* Hoboken, NJ: Wiley & Sons.

Fowler, L. 2016. Local Governments: The "Hidden Partners" of Air Quality Management. *State & Local Government Review* 48(3): 175–188.

Fowler, L. 2019. Problems, Politics, and Policy Streams in Policy Implementation. *Governance* 32(3): 403–420.

Fox, J. & G. Monette. 1992. Generalized Collinearity Diagnostics. *Journal of the American Statistical Association* 87(417): 178–183.

Konisky, D.M. & N.D. Woods. 2012. Measuring State Environmental Policy. *Review of Policy Research* 29(4): 544–569.

Koontz, T.M. 2002. State Innovation in Natural Resources Policy: Ecosystem Management on Public Forests. *State & Local Government Review* 34(3): 160–172.

Lester, J.P. 1995. Federalism and State Environmental Policy. In *Environmental Politics and Policy: Theories and Evidence*, 2nd ed., edited by J.P. Lester (pgs. 39–60). Durham, NC: Duke University Press.

Nakamura, A. & M. Nakamura. 1998. Model Specification and Endogeneity. *Journal of Econometrics* 83(1–2): 213–237.

O'Leary, R. & T. Yandle. 2000. Environmental Management at the Millennium: The Use of Environmental Dispute Revolution by State Governments. *Journal of Public Administration Research & Theory* 10(1): 137–155.

Sapat, A. 2004. Devolution and Innovation: The Adoption of State Environmental Policy Innovations by Administrative Agencies. *Public Administration Review* 64(2): 141–151.

Travis, R., J.C. Morris, & E.D. Morris. 2004. State Implementation of Federal Environmental Policy: Leveraging in the State Clean Water Revolving Fund. *Policy Studies Journal* 32(3): 461–480.

U.S. Bureau of Economic Analysis. 2019. *Regional Data* [online]. Available at https://apps.bea.gov/itable/iTable.cfm?ReqID=70&step=1 [Retrieved January 1, 2019].

US Census. 2019. *Statistical Abstract Series* [online]. Available at www.census.gov/library/publications/time-series/statistical_abstracts.html [Retrieved January 1, 2019].

U.S. Environmental Protection Agency (EPA). 2019. *TRI Basic Data Files: Calendar Year 1987–2017* [online]. Available at www.epa.gov/toxics-release-inventory-tri-program/tri-basic-data-files-calendar-years-1987-2017 [Retrieved January 1, 2019].

6

AN UNEASY PARTNERSHIP

As states serve as the ever-so-important intermediary between national society and local communities, they are the "linchpins" of the federal system. Consequently, how they fill their role in environmental protection shapes environmental federalism in significant ways. For instance, if there is a positive, cooperative national-state relationship, national policies tend to be effectively implemented, but if that relationship is negative, it is marred by conflict, policies fail to be implemented, and environmental conditions deteriorate (Scheberle, 2004, 2005). In general, when conflict takes over, environmental federalism becomes a less functional system. Troubles emerge when political incentives create a mismatch between environmental policy goals at different governmental levels, agencies are limited in their capacities to manage environmental problems, or those problems cross jurisdictional lines. For national-state relations to be positive, states must pursue an environmental policy agenda that aligns with the national government's agenda and be able to implement policies accordingly (Scheberle, 2004, 2005). Unfortunately, that scenario is not particularly common. More common are scenarios in which states push the federal agencies to adjust programs around their unique circumstances. Sometimes this is in an effort to improve environmental conditions, while other times it is more about work shirking.

While misalignment is most obvious when states prefer a less aggressive environmental agenda, it can also occur if they prefer a more aggressive agenda. The difference tends to dictate who plays the aggressor in intergovernmental conflicts, as both sides push to see their preferences for environmental protection operationalized at the state-level. In general, this means that the political will and administrative capacities of states to protect the environment are key dimensions in understanding environmental federalism as a system (Lester,

1995). Subsequently, the federal government finds itself routinely in negotiations with states over how to make this system work (Crotty, 1987; Agranoff & McGuire, 2001). In this chapter, we will look at four case studies related to CAA implementation that illustrate how different types of states interact with the national government, and some of the inherent conflicts that emerge when trying to organize a cooperative system. To reiterate our expectations from the previous chapter, first, progressive states serve as the aggressor in intergovernmental conflicts with the national government, as they push federal agencies to be more aggressive in environmental protection. Massachusetts v. EPA (2007) is a landmark case in this regard, where a group of states sued the EPA to force national regulation of GHGs.

Second, regressive states tend to be the thorn in the side of the national government as they consistently fail to comply with national standards, leading to conflict and negative interactions. In a continuation of the story of GHG regulations, Texas engaged in open warfare against the EPA in order to avoid implementing the new regulations that resulted from the Massachusetts v. EPA decision. Third, struggler states prefer to avoid conflict with the national government, and instead embrace cooperation in an effort to make up for their limited capacities. Examining North Carolina's experience with the "good neighbor" provision, we see mixed results, with the state attempting to cooperate at first but then finding itself using lawsuits to force action. Finally, delayer states also like to avoid conflicts with the national government, as they want to get by with doing the bare minimum. A prime example of this is enforcement of environmental standards in Louisiana's Cancer Alley, near Baton Rouge, in which politicians and administrators have pushed back against the EPA pursuing stronger enforcement norms, but shied away from open conflict. These four case studies highlight the complexity in national-state relationships and how internal state factors drive their character.

State as Aggressor, Part I

Massachusetts v. EPA (2007) may be one of the most famous environmental law cases to date and is often cited as the first climate change Supreme Court case. For our interest, it highlights the role of states serving as antagonists in intergovernmental conflict, and pushing federal agencies, like the EPA, to be more aggressive in their interpretation and enforcement of environmental legislation. While climate change is a central issue, this case is complex and, at its core, is a conflict over regulating air pollution. Specifically, the CAA requires the EPA to identify and regulate criteria air pollutants that "cause, or contribute to, air pollution, which may reasonably be anticipated to endanger public health or welfare" (CAA, Section 202(a)(1)). Contemporarily, this list has been limited to six pollutants (ozone, particulate matter, lead, carbon monoxide, sulfur oxides, and nitrogen oxides). These pollutants and the corresponding National

Ambient Air Quality Standards (NAAQS) are periodically reviewed based on the best science available, and the Administrator of the EPA may include additional pollutants if she deems it appropriate and establish corresponding regulations, especially if those pollutants are determined to endanger public health or welfare. However, this list has remained largely unchanged, despite petitions to include additional pollutants.

In 1999, the International Center for Technology Assessment (ICTA), a nonprofit bipartisan organization that analyzes social impacts of technologies, petitioned the EPA to include four additional air pollutants: carbon dioxide (CO_2), methane (CH_4), nitrous oxide (N_2O), and hydrofluorocarbons. These four pollutants are the core GHGs that contribute to climate change. The ICTA argued that GHGs met the criteria to be listed due to a reasonable expectation that they would endanger public health and welfare through the effects of climate change. The ICTA cited in its argument the US government's report to the United Nations Framework Convention on Climate Change, which made a similar argument. Flashforward to 2003, when then-EPA Administrator Christine Todd Whitman made the decision to deny the petition based on the grounds that the EPA did not have authority to regulate GHGs (Freeman & Vermeule, 2007). This decision was well within her duties and responsibilities as established by the CAA, but critics believed she erred in her judgment. Interestingly, in a speech two years earlier, Whitman contended that

> we must continue our efforts to stop and reverse the growth in the emission of greenhouse gases. If we fail to take the steps necessary to address the very real concern of global climate change, we put our people, our economics, and our way of life at risk.
>
> *(EPA, 2001)*

In response, three states (Massachusetts, Connecticut, and Maine) filed suit, arguing that Whitman errored in her judgment and should be more aggressive in regulating air pollutants, namely CO_2. Notably, Massachusetts was not just the leader on this case; it was also a leader on adopting state regulations that went above and beyond national standards at the time. Prior to 2003, Massachusetts had adopted more than a dozen regulatory programs to reduce CO_2, including a number of financial incentives; Connecticut and Maine had similar policies (Rabe, 2007). In other words, these states were already ahead of the national government and were pushing the EPA to catch up. Despite the overwhelming focus on climate change in this case, the core legal argument had to do with CO_2 as an air pollutant. Specifically, Massachusetts argued that the EPA had declared CO_2 as an air pollutant and had the authority to regulate it as such, so Whitman's decision was a gross failure on the EPA's part to exercise authority vested in her by the CAA (Freeman & Vermeule, 2007). Furthermore, Massachusetts argued that this failure would have a significant negative impact on its public health and welfare. In other words, the EPA was failing to exercise its duty, and states would pay the price.

Before the case made it to the Supreme Court, nine additional states joined Massachusetts's side; ten of the 12 states were either progressive or struggler states. Another ten states would side with the EPA, with seven being either regressive or delayer states (Scheberle, 2005; Rabe, 2007). While there are a few outliers (e.g., Alaska, Illinois), this is a notable division, as many of the states arguing for more aggressive national action had already established regulations in absence of national policy and were urging the EPA to realign its regulatory approach. In 2007, the Supreme Court ruled in a 5–4 decision that CO_2 and other GHGs fit the definition of air pollutants under the CAA, and, thus, the EPA had statutory authority to regulate said gases. As a result, the EPA was forced to establish new rules for the regulation of six GHGs that threatened human health and welfare. In general, this would be one of the largest expansions of CAA regulatory rules since the 1990 amendments (Freeman & Vermeule, 2007). By 2010, the EPA had provided states with guidance on how to implement these new rules, and they largely fit into the established intergovernmental regulatory framework used for other CAA regulations.

Two key questions emerge from this case. First, what happens when states are ahead of the national government in regulating emerging problems? By extension: what options do states have to force the national government to be more aggressive in these situations? States have always served as laboratories of democracy, but in order to see those experiments influence decisions at a higher level, states must have some mechanisms to force national action. This may occur in more subtle ways through vertical competitive pressures, but, ultimately, states have few coercive measures to use against national agencies. Therefore, Massachusetts and other states were left with few options to avoid seeing the progress made by their state policies wasted by free-riders in other states. On the other hand, national agencies have little incentive to be bold in policy choices when there are as many states lined up to protest as there are to support those choices. Clearly, it was in the interest of Massachusetts for the EPA to assert a more aggressive regulatory strategy, so the burden was not wholly carried at the state-level. But without resource dependencies, hierarchical authority, or coercive power, states are largely left with few ways to force national agencies into action. Consequently, we can expect these interactions to take on a much different character than those in which national and state governments are more closely aligned.

Second, how well does existing legislation help us address emerging problems? While this case deals with climate change, the legal arguments surrounded the regulation of air pollution, so there is much for us to extrapolate about how existing environmental laws apply to emerging challenges. Specifically, existing environmental legislation (e.g., CAA) will likely serve as the model for how we respond to new problems, so conflicts such as this one are emblematic of future national-state conflicts over climate policy. Nevertheless, this is a question that we will examine in more detail in a later chapter. In sum, this case conforms to our expectations for progressive states like Massachusetts (or Connecticut). The state had adopted innovative policies that were ahead of the national government. Given the opportunity, the EPA decided to punt on a controversial issue, leaving Massachusetts with few options. The state, then, decided to take the federalism issue to the courts to find some kind of relief. The result was that the national government was forced to catch up to the states; however, this case could also have gone the other way and upheld national inaction. The important point here is that states are likely to seek out aggressive ways to challenge the national government when the national government follows a policy of inaction. In general, we should expect this to be representative of the types of conflicts in which national agencies and states engage when states serve as the aggressor; that is, states seek tools to coerce national agencies into making regulatory leaps forward because more subtle mechanisms fail to do the trick.

State as Aggressor, Part II

Flashforward three years from the Massachusetts v. EPA decision. The EPA established regulatory rules for GHG permitting and required states to modify their state implementation plans (SIPs) accordingly. As the deadline approached for states to begin issuing permits, most states worked to revise their permitting processes to comply with the new federal regulations. Texas, on the other hand, drew a line in the sand and contended, "Texas has neither the authority nor the intention of interpreting, ignoring, or amending its laws in order to compel the permitting of greenhouse gas emissions" (Dawson, 2010, para. 2). Furthermore, a public statement for the Texas Commission on Environmental Quality (TCEQ) argued that "The EPA cannot measure reductions in (carbon dioxide) or any other (greenhouse gas) with this new regulation, and the EPA cannot correlate this new regulation to any environmental or health benefit" (Carlton, 2010, para. 5). In essence, the TCEQ was challenging whether the EPA should be in the business of regulating GHGs under the CAA at all. Most critics accuse Texas's state leaders of acquiescing to the demands of special interests, specifically those from the powerful oil and gas lobby. Others argue that this was a legislative process issue, as Texas state law only allowed the TCEQ to regulate air pollutants specifically identified by the state legislature. Since the state legislature was out of session (meeting every two years, rather than every year), the legislature could not amend the law, leaving the TCEQ without options.

From a broad perspective, Texas's leaders had few incentives to get on board with the EPA's new regulatory approach, nor did they have administrative capacities to devote to a new program. For instance, while the TCEQ is notorious for industry capture, it also operates in a state regulatory environment in which it competes internally against several other agencies, such as the Soil and Water Conservation Board, Department of Parks and Wildlife, the Railroad Commission (which regulates oil and gas production), and numerous independent agencies that regulate various waterways. Given the stalemate, the EPA had to act drastically, so it announced that it would instead develop a federal implementation plan (FIP). By law, the EPA is ultimately responsible for the implementation of the CAA, so if states fail to establish a SIP that complies with the national guidelines, the EPA is required to develop an FIP (Carlton, 2010). In other words, the EPA would go around the TCEQ and manage this portion of the permitting process for airborne pollutants by itself. While this also happened in a few other states, those states were working on implementing their own plans but ran into obstacles in meeting the deadline. Essentially, the EPA had no choice but to take over a portion of the state program because the state refused to manage it.

Prior to the Obama administration, the EPA only used this mechanism a handful of times in order to avoid the challenges of direct implementation. The EPA is not set up to implement national environmental programs without state cooperation, resources, or capacities, so it is impractical in many ways for the EPA to take over programs. For instance, in 1981, Idaho's state legislature, which was in conflict with the EPA, voted to defund its CAA program. After a year, the general consensus was that "the federal takeover caused more problems than it solved. EPA reportedly spent almost five times as much to maintain the Idaho program that year as the state would have spent to do the same job" (Derthick, 1987, p. 70). Based on this and similar experiences, taking over state programs was never Plan A for the EPA, but it was a necessity under certain circumstances. Specifically, an EPA spokesman said:

> EPA prefers that the state of Texas and all states remain the permitting authority . . . [but] officials in Texas have made clear . . . they have no intention of implementing this portion of the federal air permitting program.
>
> *(Carlton, 2010, para. 3)*

The EPA's FIP, of course, mimicked the national guidelines previously established and would have been overseen from the EPA Region 6 headquarters in Dallas.

It is hard to know if the EPA's plan would have been successful though, because Texas had another trick up its sleeve. Days before the EPA was set to implement its plan, the US Court of Appeals for the District of Columbia issued a stay, temporarily blocking the EPA's move to start permitting, although the US Court of Appeals for the Fifth Circuit had previously declined to do so. This, of course, left industrial facilities that produced GHGs in Texas in a unique position, as they were the only facilities in the country that were not permitted under the new federal requirements (McGowen, 2011). However, the EPA's permitting plan was only temporarily blocked, and the EPA began issuing permits in 2011 as the legal battle continued. Facilities in Texas were then left to deal with a dual permitting system, with the TCEQ issuing normal permits and the EPA issuing permits for greenhouse gases. With the DC Court of Appeals dismissing Texas's suit against the EPA, it was clear that the state was unlikely to block GHG permitting over the long-term. In response, the Texas state legislature voted to amend state law and allow the TCEQ to begin regulating GHGs and issuing permits. Subsequently, the TCEQ modified its SIP and applied to take over permitting from the EPA, which was finalized in 2014 (Marullo, 2013; Loftis, 2014).

Two key questions emerge from this case that are similar to those from the previous case. First, what happens when states are resistant to new national regulations for emerging problems? By extension, what options do national agencies have to force state governments to comply? Legislation, like the CAA, typically provides the EPA and other national agencies with mechanisms to coerce states into compliance or allows them to directly implement programs. But since national agencies are not set up to do this, it is usually not a good option. Rather, national agencies have to find ways to convince states to manage programs on their own in order for the system to function as designed. This is a difficult proposition when states actively refuse to comply with national regulations, which leads to conflicts. Of course, how these conflicts are managed has significant implications for environmental federalism, and court rulings are notable in that they establish the bounds of national and state authorities. Second, how well does existing legislation help us address emerging problems if states refuse to implement them? This may be the most important question to ask. New laws are meaningless if not implemented. If we cannot expect states to implement national regulations for environmental protection, we cannot reasonably expect those laws to effect environmental change. Again, this is a question that we will further consider in a later chapter, but this case highlights how national-state conflict can create gridlock for the implementation of national policies.

In sum, this case aligns with our expectations for a regressive state, like Texas, and makes for an important counterbalance to Massachusetts's actions in the previous case. In that case, Massachusetts sued to have the EPA be more aggressive in GHG regulation, and in this case, Texas sued to block the EPA from implementing those regulations. Policymakers in Texas and Massachusetts operate in fundamentally different contexts, so the decision calculus on the subject of GHGs is fundamentally different. In particular, this highlights the difficult position that national agencies are in when dealing with states that operate in response to dissimilar political incentives and/or administrative capacities, and the types of negative interactions that can result from the EPA attempting to play middleman between progressive and regressive states. Although this may be an extreme example, we should largely expect this to be emblematic of the types of conflicts in which the EPA and states engage. That is, some states are likely to work harder at getting out of implementing national environmental regulations than actually implementing them, which makes them a thorn in the side of the national government and leads to conflict.

State as Cooperator

Before GHG regulation became the topic de jour under the CAA, one of the most prominent issues up for debate was cross-state air pollution, or air pollution that drifted across state lines. Specifically, there was a large group of Midwestern and Southern/Southeastern states in which air pollutants were produced (i.e., upwind states) but then drifted into Northeastern states (i.e., downwind states), with those pollutants affecting both air quality and NAAQS compliance. Prior to the 1990 CAA amendments, states were left with no remedy to force other states into mitigating these effects, creating a classic free-rider problem. Essentially, upwind states produced pollution and downwind states had to deal with the consequences, both legally and environmentally. In effect, there were several states that could not regulate their way into NAAQS compliance no matter how hard they tried. However, the 1990 amendments included a "good neighbor" provision, which authorized the EPA to disapprove of a SIP if it did not effectively address interstate pollution. This meant that the EPA could determine that states were not doing enough to protect air quality in other states. The "good neighbor" provision was a major step forward in forcing states to think outside of their narrow jurisdictions when protecting the environment (Moren, 2009; Andrews, 2013).

The "good neighbor" provision also allowed states to petition the EPA to take action against identifiable sources of upwind pollution (Andrews, 2013). This provided a specific mechanism by which the EPA and states could cooperate to address a pollution problem that stretched across states. Thus, the EPA could find ways to connect actions and consequences across state lines, and states had a formal way to request help in addressing pollution sources that were outside of their jurisdictions. For those states that could not regulate their way out of noncompliance, they could then work with the EPA and their neighbors to reduce interstate pollution and give a boost to air quality. State lines create a distinct geographic boundary to any administrative capacities, so the "good neighbor" provision ultimately created an avenue for the EPA and states to leverage their capacities across borders in order to better cooperate in reducing air pollution. In other words, this allows for political incentives and administrative capacities to be realigned to meet problem parameters. After the adoption of the "good neighbor" rule, the lingering question was: how would national-state and inter-state relationships manifest under this arrangement? North Carolina's experience gives us an interesting perspective.

Due to both population growth and the subsequent development of utilities industries, air quality in North Carolina had been an issue for decades, so by the early 2000s, there was significant pressure to do something. In 2002, North Carolina passed the Clean Smokestacks Act (CSA). This "unusually creative law" (Andrews, 2013, p. 883) targeted emissions from powerplants operated by Duke Energy and Progress Energy by placing caps on total annual emissions, and led to either the modernization or retirement of 45 electric powerplants in the state over the next decade (Andrews, 2013). In 2011, the EPA accepted the CSA as part of North Carolina's SIP, which meant that it carried the weight of both national and state law. While the EPA noted at the time that there were significant improvements in air quality as a result of the CSA, those improvements are only partially attributable to emission reductions within the state (Moren, 2009; Andrews, 2013). The CSA also included a provision related to the "good neighbor" rule that directed the state attorney general to pursue legal action against any upwind pollution sources. In other words, the CSA was a two-pronged attack on air pollution that would 1) reduce emissions from electricity generators inside the state and 2) use legal mechanisms to address any emissions sources outside of the state.

North Carolina's first move on the legal front was to petition the EPA to declare powerplants in 12 upwind states as contributing to NAAQS noncompliance in North Carolina. But the EPA denied the petition, arguing that its new FIP required significant reductions in emissions across 28 Eastern states, but only in aggregate as it was based on a cap-and-trade approach rather than an approach focused on specific identifiable sources. North Carolina's attorney general then filed suit, arguing that this was an insufficient approach to remedying air pollution in North Carolina. The DC Circuit Court sided with the state and ordered the EPA to revise its plan, which would eventually lead to the more aggressive Cross-State Air Pollution Rule in 2011 (Andrews, 2013). On another front, North Carolina also sued the Tennessee Valley Authority (TVA) (a federally owned corporation), which operated power generators outside of North Carolina but near the state line. Interestingly, this lawsuit claimed emissions from TVA powerplants violated public nuisance laws in Alabama and Tennessee. A back-and-forth legal battle ensued as the case bounced between federal and state courts, but by 2011, the TVA realized it had no choice but to cooperate with North Carolina. Consequently, it acquiesced to North Carolina's demands and essentially began complying with the CSA by capping its aggregate emissions and modernizing or retiring power plants (Shogren, 2006; Smokey Mountain News, 2009; Andrews, 2013).

Two key questions emerge in this case. First, and most importantly, how do states respond when they lack the capacity to manage their own environmental problems? State leaders in North Carolina had the political incentives to clean up air quality but lacked regulatory authority over emission sources in other states. In this case, the CAA provided mechanisms to pursue collaborative action, but that did not pan out. Interestingly, even though the EPA was working to improve air quality nationally, the FIP did not align well with the goals of North Carolina, which led to conflict. Certainly, there are other case studies of the EPA and states working together well, but this case highlights how things can fall apart when the circumstances do not favor cooperation. Second, what causes cooperation to decline into conflict? Policymakers who crafted the CSA realized early on that they could not achieve their goals alone and would need to work with the EPA and other states. We can imagine that this story would have played out differently had the EPA, the TVA, and/or the states of Alabama or Tennessee agreed to work with North Carolina to reduce downwind emissions. But, when the EPA and TVA refused, it left North Carolina with few options other than to use lawsuits to force national agencies to acquiesce.

In sum, this case only somewhat aligns with our expectations for a struggler state like North Carolina. Theoretically, struggler states should have more positive interactions with national agencies, but that may not always be the case if goal conflicts emerge. In an ideal world, the EPA would have responded positively to North Carolina's original petition and created an avenue for collaboration between states to reduce regional air pollution; in the real-world, the EPA denied said petition and set off a different chain of events. While cooperation ultimately resulted, it was after a legal battle. Nevertheless, we must not forget that North Carolina favored a collaborative approach when it petitioned the EPA, but had to resort to other mechanisms when that failed. In particular, this highlights the delicate balance that exists between national and state governments, especially those struggler states that want to be aggressive in environmental protection but also need help to make those goals a reality. We should expect this to be fairly representative of the types of interactions faced by struggler states and the EPA. That is, states favor pursuing cooperation but are willing to engage in conflict (like progressive states) if they face roadblocks; thus, the politics may outweigh the administrative logic of decision-making. As such, struggler states may be some of the most pivotal states in determining whether environmental protection within the federal system is successful or not.

State as Upholder of the Status Quo

Louisiana's so-called Cancer Alley sits along the Mississippi River between Baton Rouge and New Orleans and is home to a high concentration of chemical processing plants and oil refineries. On one hand, this is a huge economic boon to the local area; on the other hand, it has also created adverse environmental conditions that are associated with negative health impacts. The latter has led to the area's unfortunate nickname due to the high rates of cancer throughout Louisiana and anecdotal evidence of cancer clusters near the plants. Notably, during the 2000s, Louisiana was in the top 15 states for toxic releases and pollution concentrations; however, this was an improvement from the 1990s, when Louisiana was among the top five states (Dubose, 2013; Lu Baum, 2019). While advocacy groups have previously opposed the presence of existing or development of new plants, environmental interest groups are not a particularly important political force in the state, and despite Louisiana being a "sportsman's paradise," there is relatively low support for environmental policy action. But Louisiana's Department of Environmental Quality (LDEQ) is relatively successful at managing environmental problems, and the EPA's only significant recommendations for improvement from SRF evaluations focus on data management practices. In other words, while there may not be much political force behind environmental protection, LDEQ administrators are very capable of complying with national guidelines.

These circumstances led to a quiet conflict between the LDEQ and the EPA in the early 2010s, when a report from the Congressional Research Service and a Louisiana-based nonprofit organization determined there was a high rate of reporting errors related to underreported accidents, spills, or excessive emissions for industrial facilities in the state (Dubose, 2013). For instance, ExxonMobil's Baton Rouge refinery was "reporting no accident to the EPA in the previous five years, while reporting 100 accidents to state agencies" (Dubose, 2013, para. 34). In response, the EPA's National Environmental Justice Advisory Council advised then-EPA Administrator Lisa Jackson to invoke the CAA's "general duty" clause, which would allow her to circumvent the EPA rulemaking process and state authority by directing plant managers on specific actions to remedy the situation. At the national-level, this set off alarm bells for Congressional Republicans, who saw this as executive overreach for a liberal cause; but it also became somewhat of a rallying call for environmental interest groups. More interesting is how Louisiana's US Senators responded. Although David Vitter and Mary Landrieu represented different parties, they found enough common ground on this issue to write a joint letter to Jackson urging her not to use the "general duty" clause. Specifically, they argued doing so would be "duplicating, confusing, and could potentially conflict with . . . current regulatory systems" (Dubose, 2013, para. 52).

Reading deeper into this case, we begin to unravel what is a rather complicated relationship between the state of Louisiana and the national government. To start, Louisiana ultimately has little interest in aggressive environmentalism, as there is scant political value to state leaders in pursuing a regulatory agenda that harms one of the state's most important industries. This would largely explain why a Republican (Vitter) and a Democrat (Landrieu) would find so much common ground on this issue. Both received campaign contributions from the chemical industry and held political positions favorable to the state's energy industry (Foley, 2014). Interestingly, Vitter would later co-sponsor the Chemical Safety for the 21st Century Act, which provided the EPA greater authority to regulate chemicals, including the preemption of state laws with federal regulations (Ballard, 2016). Noteworthy here is that it was US Senators who waded into the fight with the EPA, and not state officials. Searches of the *Times-Picayune* of New Orleans (one of the oldest and most prominent newspapers in the state) during this time period indicate that Governor Bobby Jindal made no official statement on this issue, nor did the state legislature take up any related legislation. In other words, as much as this was a challenge to the LDEQ's regulatory control, state officials seemed wary to attack Jackson or the EPA over this case (at least not publicly).

But the issue at hand comes down to two points: 1) should the EPA take a more active role in regulating Louisiana's chemical processing plants? and, 2) by doing so, would it require more effort from LDEQ to comply with national standards? It is within those points that the conflict emerges. First, the state of Louisiana may not have much interest in environmental regulation, but they also do not want to cede ground to the national government and allow the EPA to take power away from state agencies. On this point, it is very clear that Jackson's move to centralize control over environmental regulations was going to run into conflicts. President Obama's environmental policy strategy was well grounded in tactics of administrative presidencies in which the executive branch attempts to control policy through the bureaucracy and administrative actions. This would ultimately be a core conflict in environmental federalism during his administration, as states refused to give up the ground they secured in previous decades (Konisky & Woods, 2016). Second, the LDEQ is willing (and capable) of complying with EPA guidelines, but they also have a vested interest in that compliance being as easy as possible; regulated industries have a similar interest. Even if the state of Louisiana was willing to give up authority to the national government, the LDEQ would find itself in a position to jump through more hoops.

While this could have been a tipping point for the national-state balance of power, instead the conflict slowly petered out. Jackson resigned from the EPA in December 2012, citing a belief that the Obama administration would support the Keystone pipeline without taking any action under the "general duty" clause. Vitter would filibuster (more or less) the nominee to replace Jackson, Gina McCarthy, based on accusations that the EPA was being less than transparent with Congress on its rulemaking processes; his filibuster was not specifically tied to this issue (Bernstein, 2013). For all intents and purposes, this conflict ended with Jackson's resignation, and no EPA administrator since has attempted to invoke the CAA's "general duty" clause to circumvent the EPA's internal rulemaking process or state regulatory authority. Although scrutiny of the chemical processing plants in Cancer Alley continues to this day and there are still concerns about the amount of toxic releases in the area, there is no evidence to suggest that this conflict led to any policy or administrative changes for either the LDEQ or industrial facilities (Lu Baum, 2019). In essence, this case can be summarized as this: regulated facilities failed to comply with EPA standards, the EPA attempted to force facilities to do more, elected officials from Louisiana pushed back, and, at the end of the day, all remained the same.

In large part, this case conforms to our expectations for a delayer state like Louisiana. The LDEQ created a culture for minimal compliance, which lead to community concerns about environmental conditions. Since the LDEQ was essentially complying with national standards, this left the EPA with few options, as they could not directly challenge the LDEQ for work shirking or program mismanagement. Instead, the EPA attempted to expand its authority, which was challenged by Louisiana's US Senators in order to maintain the existing regulatory regime within the state. Of course, neither the EPA nor the state of Louisiana wanted to go to war, as the EPA was already under attack by Congressional Republicans and Louisiana had little to gain by doing so; as such, it never boiled over into a full-blown intergovernmental conflict. Instead, it was a quiet conflict that upheld the status quos. But this is likely emblematic of the types of conflicts in which the EPA finds itself with delayer states; that is, discreet battles that emerge when the EPA pushes for change and states find subtle ways to hinder those efforts in order to maintain existing norms.

Inherent Conflicts of National-State Cooperation

While the previous chapter provides some quantitative analysis that indicates that pollution concentrations are a function of the political and administrative contexts of states, these four case studies illustrate how that impacts intergovernmental relations and sends ripples throughout the federal system. Clearly, Massachusetts's fight with the EPA led to new conflicts with Texas; legal and policy decisions in North Carolina and Louisiana also carry significant weight for other states as they jockey for position in the federal hierarchy. Ultimately, the EPA is charged with finding a balance between progressive and regressive states, as well as between delayer and struggler states. Looking at these four case studies highlights two important things. First, states have had much different experiences with implementing the CAA, even though they are all working under the same legal constraints. That is, the political and administrative contexts have created interesting choke points in implementation that have led to conflicts, which is not surprising when one considers that some states want to see environmental conditions improve and others place the environment as a low priority (see Table 6.1 for a summary).

Second, the national-state partnership by which the CAA is implemented is wrought with conflict, as the EPA and state governments run into misalignment between their goals. Of course, these conflicts help to partially explain our findings on pollution concentrations in the previous chapter, as the character of national-state relations impacts program success. Clearly, states like Texas

TABLE 6.1 Summary of National-State Case Studies

Progressive	**Struggler**
State as Aggressor (Part I)	*State as Cooperator*
Massachusetts – state serves as aggressor to force the EPA into regulating GHGs	North Carolina – state favors cooperation with the EPA and surrounding states to improve air quality but pursues legal action when faced with roadblocks
Delayer	**Regressive**
State as Upholder of the Status Quo	*State as Aggressor (Part II)*
Louisiana – state is lax in compliance enforcement, and quiet political conflict erupts when the EPA considers changing the status quo	Texas – state tries to avoid implementing GHG permitting but acquiesces when the EPA starts issuing permits

that are sinking their limited administrative capacities into waging legal battles, and the political capital gained by elected officials in attacking the EPA, makes it nearly impossible for agencies like the TCEQ to effectively impose regulations. On the other hand, states like Massachusetts likely see the same type of legal battle with the EPA as a long-term policy strategy; they also have the administrative capacity to spare, and the political capital gained in doing so only bolsters their environmental agenda. It's the delayer and struggler states that are likely to see the biggest implications from these conflicts though, as the difference between fighting and cooperating may be the difference between success and failure. Naturally, as the complexity of environmental problems and the governing legislation grows, we can only expect that conflicts will become magnified, which is important to keep in mind as we move towards considering the implications for climate change.

References

Agranoff, R. & M. McGuire. 2001. American Federalism and the Search for Models of Management. *Public Administration Review* 61(6): 671–681.

Andrews, R.N.L. 2013. State Environmental Policy Innovations: North Carolina's Clean Smokestack Act. *Environmental Law* 43(4): 881–939.

Ballard, M. 2016. President Barack Obama Signs Chemical Safety Bill into Law. *The Advocate* [online]. Available at www.theadvocate.com. [Retrieved September, 2019].

Bernstein, L. 2013. Vitter Drops Filibuster Threat on EPA Nominee Gina McCarthy. *Washington Post* [online]. Available at www.washingtonpost.com [Retrieved September, 2019].

Carlton, J. 2010. EPA to Take Over Greenhouse Gas Permits in Texas. *Associated Press* [online]. Available at apnews.com [Retrieved September, 2019].

Crotty, P.M. 1987. The New Federalism Game: Primacy Implementation of Environmental Policy. *Publius* 17(2): 53–67.

Dawson, B. 2010. Texas Officials Say They Won't Implement EPA's Climate Rules. *Texas Climate News* [online]. Available at texasclimatenews.org [Retrieved September, 2010].

Derthick, M. 1987. American Federalism: Madison's Middle in the 1980s. *Public Administration Review* 47(1): 66–74.

Dubose, L. 2013. The EPA Can Clean the Air in "Cancer Alley," Louisiana. *Washington Spectator* [online]. Available at www.washingtonspectator.org [Retrieved May, 2016].

Foley, L. 2014. Chemical Industry Political Spending Surges. *Environmental Working Group*. Available at www.ewg.org [Retrieved May, 2016].

Freeman, J. & A. Vermeule. 2007. Massachusetts v. EPA: From Politics to Expertise. *The Supreme Court Review*: 51–110.

Konisky, D.M. & N.D. Woods. 2016. Environmental Policy, Federalism, and the Obama Presidency. *Publius* 46(3): 366–391.

Lester, J.P. 1995. Federalism and State Environmental Policy. In *Environmental Politics and Policy: Theories and Evidence*, 2nd ed., edited by J.P. Lester (pgs. 39–60). Durham, NC: Duke University Press.

Loftis, R. 2014. Texas Oks Taking over Greenhouse Gas Permits from EPA. *Dallas Morning News* [online]. Available at dallasnews.com [Retrieved September, 2019].

Lu Baum, J. 2019. They Don't Call It "Cancer Alley" for Nothing. *Big Easy Magazine* [online]. Available at www.bigeasymagazine.com [Retrieved September, 2019].

Marullo, J. 2013. New Texas Legislation Authorizes TCEQ to Permit Greenhouse Gas Emissions. *The Energy Law Blog* [online]. Available at theenergylawblog.com [Retrieved September, 2019].

Massachusetts v. Environmental Protection Agency. 2007. Supreme Court of the United States. 549 U.S. 497.

McGowen, E. 2011. Court Backs Texas Revolt Against EPA's New Greenhouse Gas Rules. *The Guardian* [online]. Available at theguardian.com [Retrieved September, 2019].

Moren, H. 2009. The Difficulty of Fencing in Interstate Emissions: EPA's Clean Air Interstate Rule Fails to Make Good Neighbors. *Ecology Law Quarterly* 36(2): 525–552.

Rabe, B. 2007. Environmental Policy and the Bush Era: The Collison between the Administrative Presidency and State Experimentation. *Publius* 37(3): 413–431.

Scheberle, D. 2004. *Federalism and Environmental Policy: Trust and the Politics of Implementation*, 2nd ed. Washington, DC: Georgetown University Press.

Scheberle, D. 2005. The Evolving Matrix of Environmental Federalism and Intergovernmental Relationships. *Publius* 35(1): 69–86.

Shogren, E. 2006. North Carolina Sues TVA to Clean Up Pollution. *NPR* [online]. Available at www.npr.org [Retrieved September, 2019].

Smokey Mountain News. 2009. *Landmark Victory: NC Forces TVA to Clean Up Its Act* [online]. Available at www.smokymountainnews.com [Retrieved September, 2019].

U.S. Environmental Protection Agency (EPA). 2001. *Remarks of Governor Christine Todd Whitman, Administrator, United State Environmental Protection Agency at the G8 Environmental Ministerial Meeting Working Session on Climate Change* [online]. Available at https://archive. epa.gov/epapages/newsroom_archive/speeches/ef9a58127adb3b4b8525701a0052e348. html

7
THE TANGLED WEB OF LOCAL GOVERNMENT

As the intermediary in the federal system, states bridge the gap between national society and local communities. The previous chapter examined some of the inherent conflicts that emerge when dealing with the former; this chapter makes a similar examination in regards to the latter. While national-state interactions have been the focal point for federalism scholars (and environmental federalism in particular), many argue that the state-local dimension is the new battleground (Riverstone-Newell, 2012; Fowler & Witt, 2019). More specifically, national-state conflicts are more or less continuing patterns that have existed for the last half century; on the other hand, local governments have found new footing in the federal system by creating innovative policies that challenge state leadership and move beyond myopic policy regimes established by national or state authorities. Similar to the national-state dimension, if state and local governments work together to match resources and institutions with problem parameters, environmental policies can effectively be implemented to both achieve national goals and meet local needs. Unfortunately (again), that scenario is relatively uncommon, and local governments end up playing a variety of roles both inside and outside of state environmental regulatory regimes. Of course, the nature of their role shapes their interactions with states and capacities to contribute positively to environmental protection.

In this chapter, we will look at four case studies related to the protection of water quality, as well as CWA implementation, that illustrate the variable roles local governments play, as well as some of the inherent conflicts that emerge. Although local governments are a diverse group, we believe these cases highlight how factors occurring at the state-level shape interactions with local governments and their roles in the overall environmental protection scheme. To reiterate our expectations from Chapter 5, first, progressive states dominant environmental protection so much that local governments are left with little wiggle room; in effect, they end up essentially serving as subunits of state agencies. California provides an interesting example of this, where sophisticated state institutions create a rigid system that turns local governments into compliance managers. Second, given the tendency of struggler states to cooperate with any and all in order to expand their administrative capacities, local governments are largely treated as partners in environmental protection and relationships take on a collaborative character. These trends are apparent in Florida, where state agencies use partnerships to leverage local implementation resources and social capital to achieve shared policy goals.

Third, in delayer states, state agencies must find ways to create politically feasible solutions to environmental problems that also meet national standards. That is, the challenge is political rather than administrative, as states have the capacities but environmental protection is not always politically acceptable. West Virginia demonstrates this by utilizing community groups to develop mechanisms at the local-level that garner stakeholder buy-in. Finally, in regressive states, local governments mimic the behavior of state governments in which they retreat from environmental protection. In large part, this is because there are few political incentives, and local capacities are simply not strong enough to overcome the lack of state engagement. These behaviors are highlighted by Ohio, where minimal state effort has left local governments in a vulnerable position. Before we undertake these case studies though, we think it is prudent to provide a brief background on local governments and their rising role in environmental policy.

The Rise of Local Environmental Policy

While the number of states has remained stable at 50 since Alaska and Hawaii were admitted to the Union in 1959 and there has been a single national government since 1776, the number of local governments has fluctuated over time, ranging from about 84,000 in 1992 to about 90,000 in 2012. When we remove school districts and special districts, there are approximately 40,000 general purpose governments (i.e., counties, municipalities, townships) with municipalities making up the lion's share and serving as the key functionary for providing public services to a mass of the US population at the local-level. Of course, those numbers vary widely by state, with Hawaii having a single municipal government (i.e., Honolulu) and Pennsylvania, Texas, and Illinois having more than 1,000 each. Furthermore, cities serving populations of over 100,000 amount to only about 300 in 45 states (Census, 2019). Suffice it to say, it is difficult to make sweeping statements about this diverse group of governments that faces different legal, institutional, political, and socioeconomic realities (Langan & McFarland, 2017). Nevertheless, previous scholars provide us with a few general ideas and expectations about what drives their behaviors and how they fit into the intergovernmental system.

For better or worse, authority in the federal system tends to be allocated based on political purposes, rather than on efficiency, effectiveness, or democracy (Wood & Bohte, 2004; Volden, 2005). This may be no more apparent than in looking at the state-local balance of power. In general, scholars argue that local policy innovation and, by extension, its role in the intergovernmental system is driven by both top-down and bottom-up elements (Bowman & Kearney, 2011; Fowler & Jones, 2019). In terms of top-down forces, states have routinely used their hierarchical power in the federal system to put constraints on local governments via resource dependencies, legal mechanisms, or political coercion (Feiock & Scholz, 2009; Terman & Feiock, 2015). Of course, this is much easier under the Dillon's Rule doctrine, where local governments are seen as creations of the state and must be granted authority to act, which creates rather narrow areas of policy discretion. This is in comparison to the Home Rule doctrine in which local governments are seen as independent of states and enjoy broad discretion, but states can use their authority to create limitations to that discretion. Currently, around ten states operate primarily under the Home Rule doctrine, but several states, such as Colorado, treat some classes of local governments as Home Rule and others as Dillon's Rule entities (Richardson, 2011; Hicks, Weissert, Swanson, & Bulman-Pozen, 2018).

Despite the uptick in preemption legislation by conservative state legislatures attempting to stop progressive cities from creating vertical political challenges through innovative policies in recent years (Riverstone-Newell, 2012, 2013; Fowler & Witt, 2019), state-local relations in environmental policy tend to be more heavily influenced by the use of second-order devolution (i.e., state to local devolution, as opposed to national to state, which is first-order) (Woods & Potoski, 2010; Fowler & Jones, 2019). Specifically, some states have followed the example set at the national-level by pushing primary responsibility for managing the CAA or CWA to local governments, which then take on day-to-day functions while state agencies provide oversight. States have found good administrative and political reasons for doing so. Administratively, local governments have acute expertise and capacities that can be deployed in their jurisdictions to supplement state efforts; politically, it provides state elected officials an opportunity to shift the blame for unpopular programs and to claim credit for any successes (Agranoff & McGuire, 2001; Volden, 2005; Krueger & Bernick, 2010; Woods & Potoski, 2010; Reed, 2014; Fowler & Jones, 2019). Notably, this has bred different forms of intergovernmental management, such as the top-down approach in which local governments are strictly directed by states, and the donor-recipient form in which local governments are treated as partners and provided opportunities to negotiate their roles and responsibilities (Agranoff & McGuire, 2001).

On the other hand, scholars are also observing more and more bottom-up behavior from local governments, as they challenge state domination of policy agendas for air quality (Fowler, 2016, 2018; Fowler & Jones, 2019; Fowler & Rabinowitz, 2019), watershed management (Lubell & Fulton, 2008), energy

(Byrne, Hughes, Rickerson, & Kurdgelashvili, 2007; Davis & Hoffer, 2010; Davis, 2014), and climate change (Krause, 2011, 2013), among other social, economic, and public safety issues (Riverstone-Newell, 2012, 2013). Again, there are both administrative and political reasoning for this. Administratively, local managers are responding to the complex policy problem they face, which is exacerbated by institutional collective action dilemmas and defective state regulatory regimes. That is, they recognize a failure in the federal institutions surrounding them and respond with unconventional tactics, such as forming regional governance networks or interlocal partnerships (Agranoff & McGuire, 1998, 2001; Chen & Thurmaier, 2009; Feiock & Scholz, 2009). Politically, local elected officials see an opportunity in vertical political competition and creating alternative policy venues that may attract new resources or bring prestige to their community (Volden, 2005; Shipan & Volden, 2006, 2008; Riverstone-Newell, 2012, 2013). Interestingly, there is some debate about whether these trends are driven more by opportunism or by interest in pursuing shared policy goals (Conlan, 2006; McGuire, 2006). In either case, we should expect local governments to respond to institutional failures with policy innovations and challenges to state leadership.

Given the competing trends, Fowler and Jones (2019) observe that local agencies operating in the same environmental policy space may in fact function in very different ways, and identify three types of agencies: fully devolved agencies, which have the authority to both set environmental quality standards and enforce federal or state standards; state administrative subunits, which have only the authority to enforce federal or state standards; and activist agencies, which have neither authority and operate outside of state-organized environmental regimes. Notably, the authority to set standards is chiefly tied to political decision-making about the balancing of interests, while enforcing of standards is tied to administration (Potoski & woods, 2002; Woods & Potoski, 2010; Fowler & Jones, 2019). Thus, the delegation of authorities aligns with our general framework surrounding the political versus administrative dimensions of environmental policy and how states shift responsibilities to local governments. Furthermore, in a related study, Fowler (2020) indicates that which type of local agencies states incorporate into their policy implementation has significant impacts in environmental outcomes. In sum, state-local relations are a complex issue and have nontrivial impacts on the success of national environmental policies.

State Domination, Local Compliance Management

The state of California runs one of the most sophisticated environmental policy regimes in the US, via multi-level governance institutions; however, it leaves only a limited role for local governments, which largely serve as compliance managers, as opposed to implementation partners or policy innovators.

Constitutionally, California is chiefly a Home Rule state in which cities are allotted broad authorities to make policies in response to local interests, and do not require specific authorization from the state to do so unless the state has adopted specific laws preempting or restricting local action; however, non-charter cities are governed by Dillon's Rule. In practice, court rulings have generally found that local governments cannot pass ordinances that conflict with state laws, which is a difficult proposition in a state that adopts aggressive and rigorous policies, particularly for environmental protection (League of California Cities, 2019). For instance, in 2000, when a charter city passed an ordinance to eliminate fluoridation of the city's water supply, the attorney general opined "that the ordinance conflicted with state law . . . [and determined that] a state statute requires local entities to accept state funds and install fluoridation systems" (League of California Cities, 2019, p. 8). In other words, local governments have little leeway and are more or less consigned to compliance management (even if local leaders are opposed to specific policies), as long as the state provides resources and guidance to support that role.

At the broadest level, the California Environmental Protection Agency (CEPA) is charged with providing environmental protection and is the lead agency for implementing national and state policies. Typical of most state environmental agencies, parts of that mission are subdivided under CEPA departments or divisions (e.g., Department of Toxic Substances Control) (CEPA, 2019). Adding further nuance to this organizational scheme are semi-independent state control boards that take the lead in specific environmental mediums (e.g., Water Resources Control Board), which are designed to provide additional policymaking venues and incorporate substantive expertise. State boards are then supplemented by a series of regional control boards that oversee and adapt policies to fit specific geographic areas. Both state and regional control boards are led by gubernatorial appointees (confirmed by the State Senate) and are part of CEPA. Specifically for the CWA, the California State Water Resources Control Board and the associated regional water resources control boards are the lead agencies for implementation (California Water Boards, 2013, 2019; Nickles, 2013). For all intents and purposes, regional control boards function as state administrative units, and local governments are responsible for complying with any parameters they set. Consequently, regional control boards provide local expertise and create a policymaking venue to address local interests, both functions that cities or counties would otherwise serve, which leaves little room for them in this policy regime.

Two specific examples highlight these trends. In 1999, the EPA finalized new rules for the National Pollutant Discharge Elimination System (NPDES) Phase II Stormwater Program. This new rule was meant to address growing concerns surrounding nonpoint source pollutant discharges, as opposed to point sources regulation, which is the chief vehicle used by the CWA for water quality improvement. Under the new rule, state governments were to adapt their

CWA implementation plans and provide guidance to local governments, which would then implement new programs based on local circumstances (White & Boswell, 2006). However, in analyzing responses, White and Boswell (2006) found that local governments were more motivated by simple compliance with federal and state mandates than developing quality responses to reducing non-point pollutant discharges. Notably, local governments were also reluctant to invest their own resources into the new program or develop plans without specific guidance on implementation requirements. Local managers rated implementation ease as a key criterion in their policy and planning development processes and organized evaluation tools around the existing stormwater activities that allowed them to take credit for efforts already in place. On the other hand, performance did improve when model programs were made available by regional boards. Taken together, this would generally suggest that local managers in California largely see themselves as compliance managers rather than implementation partners or policy innovators.

Although efforts at creating integrated regional water management (IRWM) systems have been ongoing in California since the 1980s, a new iteration was kicked off by the passage of Proposition 50 in 2002, which dedicated $3.44 billion for water protection projects. Generally, IRWM systems are designed to increase cooperation and remove institutional barriers to collective action, and they are typically held in contrast to fragmented regulatory regimes that represent status quo water politics in many Western states (Lubell & Lippert, 2011). In examining the IRWM in the San Francisco Bay Area, Lubell & Lippert (2011) found that local governments were generally less participatory in the new system as compared to the regional water control boards, environmental special districts, and non-governmental organizations (NGOs) in the area. Although local government managers indicated that they thought IRWM was a better way to achieve water management goals, maintenance of the existing status quo proved to be a barrier to collaborating with other policy actors. The state's efforts to break away from fragmented regulation have seen incremental success, which highlights both the state's role in setting the pace for environmental protection and the resistance to local governments doing more than compliance management (Lubell & Lippert, 2011).

This case begs two interesting questions. First, do sophisticated policy regimes at the state-level box local governments out of environmental protection? Certainly, in California, the state-led policy regime supplants local governments with regional control boards. Considering that those regional boards bring local expertise and policymaking venues into state institutional structures, the added utility of incorporating local governments into the state-led policy regime is somewhat limited. Furthermore, legal requirements to comply with and implement state environmental regulations strain local capacities that may otherwise go towards policy innovation. As such, local governments likely have little incentive or ability to overcome the institutional barriers to playing a more active role in environmental protection. While cities and counties remain important entities in protecting water quality, they are ostensibly compliance managers in California, as opposed to implementation partners or policy innovators like they are in other states. Given that the state of California has both the political incentives and administrative capacities to provide environmental protection at a high level, it should be no surprise that control over this policy area has been centralized at the state-level and a quasi-comprehensive (i.e., connecting state to local within a single organization) institutional structure has been developed that leaves local governments on the sidelines.

Second, is there a parallel between this state-local balance of power and that of the national-state balance? There are important similarities between the EPA's top-down approach used during the 1970s and California's current approach. In both cases, power was centralized and lower governmental units functioned as compliance managers. The national-state balance of power evolved over time; on the other hand, in California, the state-local balance seems to have stabilized. This begs further questions about how powers are distributed across three federal levels and whether centralization at any single level is efficient. In general, this case aligns with our expectations for progressive states, although state policy regimes in other progressive states may not be as sophisticated. In sum, the state sufficiently dominates environmental protection and relegates local governments to compliance management roles. California designed a sophisticated multi-level institution that connected policy implementation efforts with local expertise and policymaking venues controlled by the state. In turn, this leaves little room for local governments to adapt national or state policies to unique local sociopolitical, economic, or technical challenges, as regional control boards already fill that function. The important takeaway is that if states dominate environmental protection by developing policy regimes that centralize control, local governments will have only a limited role. We should expect this to be representative of the balance of state-local roles in environmental protection when states have both the political incentives and administrative capacities to dominate.

State Collaboration, Local Partners

The state of Florida has organized environmental protection in a way that places much more emphasis on developing partnerships with other policy actors, which positions local governments to function as implementation partners, as opposed to compliance managers. Like California, Florida is a Home Rule state; unlike California, Florida has not created de facto preemption of local environmental protection through a series of state laws (Florida League of Cities, 2011). That is, although local ordinances cannot create conflicts, Florida's state environmental laws tend to not be overly rigorous in their requirements and allow for discretion at the local-level. Additionally, while the Florida

Department of Environmental Protection (FDEP), the lead agency for implementing national and state environmental laws, employs regional offices that review permits and conduct compliance monitoring, those regional offices do not serve as policymaking venues for local interests (FDEP, 2019). Furthermore, as the FDEP faces practical limitations in pursuing its mission (e.g., limited resources), the state also leans on local governments as partners in order to maintain political coalitions, supply additional implementation capacity, and help legitimize administrative decisions (i.e., provide operational localism) (Reed, 2014). Consequently, this has generally left local governments with a much more active role in environmental protection.

One example of this is the Suwannee River Partnership in Florida's panhandle. Historically, the Suwannee River has exceeded water quality standards for nitrite, which has resulted from nonpoint sources (i.e., agricultural runoff) in the area and presented a challenge for more conventional command-and-control regulatory approaches. In 1999, the Suwannee River Partnership was established via a Memorandum of Understanding between 42 state, local, and national agencies and agriculturally focused NGOs (e.g., Florida Farm Bureau). The purpose was to cooperatively implement conservation programs to improve environmental quality throughout the watershed (Lubell, 2004; Dedekorkut, 2005). In order to do so, the Partnership both supported regional-level conservation activities and encouraged farmers to employ on-site management practices that reduced pollution run-off into surrounding waterways. Of course, farmers' use of these best management practices was a key criterion for success. To this end, Lubell (2004) finds that trust in local governments served as one of the most important indicators in farmers' willingness to participate; on the other hand, trust in the FDEP, the EPA, and government in general were not substantive predictors. In other words, the success of the Suwannee River Partnership has hinged considerably on the social capital built by local governments that the FDEP was then able to leverage in order to encourage farmer participation in conservation programs.

Collaborative water management schemes, such as the Suwannee River Partnership, have been increasingly popular across Florida since the 1990s as urban development has threatened ecologically sensitive areas (e.g., the Everglades) and the state has grappled with how to effectively balance competing environmental and economic interests. Scholars have partially attributed the success of these schemes to contributions from local governments through both formal (e.g., planning capacities) and informal (e.g., information sharing, legitimization) mechanisms (Brody, Highfield, & Carrasco, 2004; Yates, Stein, & Wyman, 2010). Local agencies have also played the role of guinea pigs via adaptive management and have experimented with best management practices as they test different ways to achieve shared policy goals (Borisova, Racevskis, & Kipp, 2012). Still, these functions require some external institutional trigger, such as state directives or resources, in order to break local policy communities out of their status quos (Gerlak & Heikkila, 2007). Consequently, the FDEP is not a simple bystander, watching as local agencies unilaterally pursue innovative solutions to environmental problems. Rather, the FDEP, as well as other state agencies, help create regional cooperative networks and institutional mechanisms to stretch local administrative capacities beyond their narrow jurisdictions. Subsequently, the implementer partner moniker sums up how state and local agencies work together to address water quality issues in Florida.

Looking at Florida's experience with water quality management leads us to three important questions: 1) how much should states rely on local governments to protect the environment? 2) how does managing through intergovernmental collaboration differ from managing through coercion? and 3) is there a parallel here with national-state relations? Given the limited administrative capacities at the state-level, the FDEP relies on local governments to create operational localism for its policies. In turn, this allows local governments to influence how environmental protection works within their communities. There are clearly pros to this arrangement for both state and local governments if both sides are largely in agreement about their mission. However, this may lead to notable conflicts if disagreements emerge or local governments are unwilling to play the role of partner. This system also opens the FDEP up to institutional failure, as it does not have effective control over the entire environmental protection system. As such, the FDEP and other state agencies have to employ collaborative approaches (i.e., shared policy goals, bargaining) to intergovernmental management in order to cajole local governments to follow their lead, rather than resort to coercion (i.e., resource dependencies, preemption) to force compliance. On one hand, this removes powerful tools from the state; on the other hand, it likely leads to a cooperative culture where partners work together to pursue opportunistic agendas.

Certainly, there are similarities here with the post-Reagan approach to environmental federalism at the national-level, where states became more like implementation partners than compliance managers. Our case study does not begin to examine the potential conflicts this may lead to though, and previous chapters note that at the national-level, not all states have responded positively, leading to trouble. Further examination of Florida's cities and counties would likely uncover local resistance to the state's environmental agenda and present additional questions about the efficacy of using local governments as implementation partners. Nevertheless, our case study highlights how a lack of administrative capacities in struggler states leads to a more collaborative approach to working with local governments, as they become implementation partners. In sum, the state leaves flexibility in its environmental protection scheme in order for local governments to match policies to unique challenges in their communities. Additionally, this allows local governments to leverage their social capital to encourage compliance from other policy actors. Florida has utilized this strategy through support of collaborative watershed management efforts at the community-level. While this approach may create its own unique challenges and risks, it does allow governmental units to organize themselves around shared policy goals and move away from myopic top-down hierarchies. We should expect this to be generally representative of state-local relations in environmental policy when states have the political incentives but not the administrative capacities to unilaterality accomplish their mission.

State Status Quos, Community Political Feasibility

Although West Virginia is not known as an environmentally friendly place, it has organized state-level institutions to meet national standards and then allowed flexibility for innovation at the local-level. Consequently, municipalities and community associations serve as important components of the environmental protection scheme. West Virginia's long history with coal mining has led to environmental problems and conflicts, as the coal industry has been placed on a pedestal as a cultural institution that is central to community economic identities across the state (Bell & York, 2010). This has created an interesting context for environmental management. On one hand, national-state conflict over meeting environmental standards stretching back to the 1970s has forced state agencies to develop the requisite capacities to comply with EPA guidelines, even though the state's political leaders are still heavily influenced by polluting industries. By upholding the status quo in CWA implementation, it allows West Virginia to avoid conflicts with the EPA and environmental advocacy groups that closely monitor industries within the state. On the other hand, coal mining has created a litany of environmental problems that negatively impact citizens and communities and for which local governments may be motivated to address (Bell & York, 2010). In order to balance these competing challenges, state agencies cast themselves as compliance managers for national policies while allowing for community-based innovations and adaptation.

Historically, West Virginia has been a Dillon's Rule state, in which local governments were allotted only narrow policymaking authority as specifically outlined by the state legislature, and scholars previously ranked West Virginia local governments as having among the least autonomy in the US (Wolman, McManmon, Bell, & Brunori, 2008). In response to the Great Recession in the mid-2000s, municipalities began petitioning the state legislature for Home Rule to provide them with more flexibility in dealing with financial constraints during economic downturns (Webb, 2012). In 2007, the state launched a pilot program to provide four cities (Bridgeport, Charleston, Huntington, and Wheeling) with Home Rule status as an experiment in local innovation, and by 2019, 34 municipalities had been granted permanent Home Rule status. The goal of this experiment was to create an "incubator of innovation" by giving municipalities "the ability to implement ordinances, resolutions, rules and regulations that fit their specific dynamics" (West Virginia Municipal League, 2019). This highlights a larger trend where municipalities have been recast as policy innovators, compared to their previous role as compliance managers, as the state tries to shift more responsibility for solving community problems to the local-level.

Of course, these trends extend to environmental policy as well. For instance, the West Virginia Department of Environmental Protection (WVDEP) has been noted for best practices for creating templates for both inspections and penalty assessments, and it provides staff with an enforcement handbook for determining appropriate actions in response to violations (EPA, 2019d). Additionally, while the state does not have a formally recognized environmental dispute resolution program, its strategy is generally to negotiate with violators before going to court, in order to find resolution without a legal battle (O'Leary & Yandle, 2000). In general, this would indicate that the WVDEP approaches policy implementation as an exercise in compliance management and tries to avoid conflict when possible. Although municipalities in the state play no formal role in CWA implementation, there is evidence of a few interesting innovations that have empowered community-based innovation. Similar to other states, groups of stakeholders have formed collaborative watershed management groups to bring together government, industry, environmentalists, and the general public (Cline & Collins, 2003; Moore & Koontz, 2003). Notably, scholars indicate that implementation of Total Maximum Daily Loads (TMDLs) tends to be more successful if stakeholder groups are present, and differences in implementation success between West Virginia and Ohio, its neighbor, may be partially attributable to the grassroots character of West Virginia groups, as well as the WVDEP's centralized regulatory oversight (Cline & Collins, 2003; Hoornbeek, Hansen, Ringquist, & Carlson, 2008, 2013).

Furthermore, Benham, Zeckoski, and Yagow (2008) attribute the success of TMDL implementation on the North Fork of the South Branch of the Potomac River to a combination of state-level support and community-based engagement. For instance, the WVDEP provided funding and technical assistance as well as a watershed model in order to inform and engage stakeholders, who then participated in the development of work and implementation plans. While the decisions on TMDLs were ultimately approved at the WVDEP's central office in Charleston, much of the development work involved community-based stakeholders (Hoornbeek, Hansen, Ringquist, & Carlson, 2008). Most indicative of West Virginia's willingness to push decision-making to the local-level though is the water quality trading system on the Cheat River, which served as one of the original EPA pilot projects beginning in 2002. The water quality trading system was designed to "facilitate clean-up of sites by providing cost-effective reduction credits to other regulated sources facing more stringent water quality effluent limits under the Cheat TMDL" (Breetz et al., 2004, p. 335). The original working group, the Cheat Trading Stakeholder Group, developed the trading program, which was then incorporated into West Virginia's NPDES regulations. In other words, West Virginia provided stakeholders at the community-level with the discretion to develop a program to meet water quality standards based on the unique issues occurring within their communities (Breetz et al., 2004).

For the most part, this illustrates that the WVDEP largely maintains control over water quality management but is willing to devolve decision-making authority to the community-level in order to allow stakeholders to develop politically feasible mechanisms to address environmental problems. Of course, this suggests two overarching questions. First, are local governments the only mechanism for connecting state policy regimes to communities? As West Virginia demonstrates, the answer is no. Although local governments can fulfill this function, West Virginia has not leaned as much on local governments as they have on community-based stakeholder groups, of which local governments are a part. The difference is that local governments are not serving as parts of the overall institutional mechanisms by which the CWA is implemented; rather they are providing input as part of a larger group looking at how those institutional mechanisms should function. Importantly, "better" in this instance is likely tied to what is politically feasible rather than what is efficient or effective. In other words, the WVDEP does not have an administrative capacity problem as much as it has a problem with the political acceptability of environmental protection. Consequently, engaging stakeholders and communities helps to cultivate local social capital more than it allows the WVDEP to access implementation capacities.

Second, how does West Virginia contrast with Florida's use of local governments as implementation partners? As alluded to earlier, Florida relies on local governments to supplement its administrative capacities, while West Virginia relies on community stakeholders to provide the political legitimacy to make programs work. This is important in that local collaborative activities may look similar, but how the state engages with and utilizes local policy actors is for different purposes. As such, this case study provides an important contrast to the previous one, as well as aligns with our expectations for a delayer state. In sum, the state employs its administrative capacities to see that CWA implementation effectively meets national standards in order to avoid external conflicts. Additionally, it leverages community-based groups to develop politically feasible solutions that stave off internal political conflicts. West Virginia has utilized this approach through incorporating stakeholders at the local-level into statewide efforts. Although there is little expectation that West Virginia (or other delayer states) will exceed national environmental quality standards with this approach, it does allow them to balance the competing interests at national- and community-levels and to avoid conflicts with national or local governments. We should expect this to be generally representative of state-local relations when states have the administrative capacities but not the political incentives to make large-scale commitments to environmental protection.

State Retreat, Local Struggles

Although Ohio was at the epicenter of state environmental activism in the 1970s, economic shifts have pushed environmental issues to the backburner for the state and left local governments in a vulnerable position. Although the Ohio Environmental Protection Agency (OEPA) was one of the first stand-alone pollution prevention agencies in the US, Ohio has suffered from deindustrialization, urban decay, and retrenchment as manufacturing industries that were once economic anchors have declined since the 1980s (Bowen, 2014). Manufacturing industries have also left a legacy of environmental problems with which the state is still grappling. Despite this, there remains a disinterest in prioritizing environmental protection above economic development, as politically powerful lobbies remain influential and workers cling to hopes of a resurgence of traditional occupations. Consequently, politicians tend to be wary of environmental commitments, which leads to conflicts with the EPA and advocacy groups as state leaders fail to effectively address lingering environmental problems, and the OEPA has long been accused of industry capture (Scheberle, 2004; Pautz, 2010). Within this framework, local governments have been left to fend for themselves as the state has largely retreated from environmental protection.

One striking indicator of this trend is the number of consent decrees between the EPA and local governments related to water quality in Ohio. Consent decrees are legal settlements and, in environmental law cases, tend to involve an agreement to restructure policies in order to meet environmental standards. In many cases, the EPA brings forth legal actions after it determines a systematic failure to comply with national environmental laws, and the courts serve as supervisors to ensure that the terms of the agreements are met. When consent decrees are entered into by local governments and the EPA, local governments do not necessarily admit fault, but they are being forced through legal action to alter their environmental management practices (Percival, 1987; O'Leary, 1990). While states are typically plaintiffs alongside the EPA in these cases, putting them on opposing sides with local governments, the EPA is the lead plaintiff and lawsuits are handled at the federal-level (EPA, 2019c). Notably, the OEPA could provide additional funding and support to municipalities or take over operations of municipal wastewater systems, but that would require the state to be proactive in environmental protection. Thus, if the EPA is negotiating settlements, it is a sign that the state is essentially choosing to "punt" and have the EPA force alternative management practices, rather than Ohio using its own policy tools.

Of the EPA's 38 CWA settlements with local governments between 2000 and 2010, six were from Ohio. The remaining 26 are distributed across 17 different states, Washington, DC, and Puerto Rico; Indiana, Kentucky, and Maryland are the only other states in which more than two originated (EPA, 2019c). In other words, Ohio has the highest concentration of CWA compliance violations by local governments that elicit EPA legal action in the nation. In almost all of these cases, municipal sewage systems were found to be in violation of NPDES regulations and were dumping excess wastewater into public waterways. At a minimum, settlements include a commitment on behalf of defendants to bring sewage systems into compliance, but some also involve civil penalties. For instance, in 2002, the City of Youngstown agreed to spend $12 million on short-term improvements over six years and $100 million over two decades as part of a long-term plan, and, in 2009, the City of Ironton agreed to a plan that would separate its stormwater and sanitary sewers by 2026 and paid a $98,000 fine (EPA, 2019a, 2019b). Notably, the state of Ohio plays a minimal role in these agreements, as the EPA leads settlement negotiations, as well as provides oversight and enforcement, and federal courts have legal jurisdiction. In some cases, the OEPA may serve as an advisor or provide oversight, but rarely does it provide additional support or resources to local governments to facilitate a return to compliance (EPA, 2019c).

Lack of support for local water management efforts in Ohio extends to other areas as well. For instance, beginning in 1997, Ohio has pursued collaborative regional watershed management, similar to Florida and West Virginia (Ruhl, 1999). These efforts were meant to be bottom-up and the state has provided guidance, but it has left organizing regional cooperative networks and planning processes to local actors. Financial support has been limited and typically requires collaborative groups to develop action plans first (Ruhl, 1999). In examining these efforts, Koontz and Newig (2014) found that participants "described implementation challenges due to a lack of resources" as it related to both coordination activities and technical tasks to improve water quality, which was exacerbated when the state cut funding to soil and water conservation districts. A lack of support has created further implementation challenges, where local opposition to projects and pushback from landowners have been barriers to progress (Koontz & Newig, 2014). Additionally, the dominant stakeholder groups, such as electric utilities and municipal wastewater treatment plants, prefer to focus on pro-growth strategies; conversely, other coalitions advocate for more action on nonpoint sources (Maddock, 2004). While the state may find an advantage in pushing for bottom-up efforts to improve water quality, local governments have generally lacked the resources and social capital to implement action plans, although many participants believe that collaboration has had a positive impact (Koontz & Newig, 2014).

The most pressing question that this case suggests is: can local governments protect the environment without state support? Or, as a corollary, what happens when states retreat from environmental protection? The answer to the first question is likely no. In general, there are few local governments with the resources, authorities, or capacities to effectively protect the environment, so at a minimum they are reliant on states to make up for their inherent limitations. Of course, this is complicated by the fact that if states have primacy, they are legally responsible for managing national environmental laws, such as the CWA. Therefore, even if cities like Youngstown wanted to ramp up their efforts, there is a distinct institutional barrier formed by the lack of action at the state-level. Although many local governments have been motivated to be innovative by those circumstances, previous research generally shows that local efforts are only successful if they are coordinated with and complement state efforts (Fowler, 2019, 2020). In other words, state and local governments have different strengths and weaknesses when it comes to environmental protection, and it is unlikely that local governments would be able to completely replace states. Consequently, the success of the overall system may be dependent on both state and local governments.

On the other hand, if states retreat from environmental protection, or to a lesser extent, do the bare minimum, it puts local governments in a vulnerable position where they are on the frontlines of environmental protection but without the institutional support to be successful. This case highlights these issues, as well as aligns with our expectations for regressive states. In sum, the state does the bare minimum in environmental protection, pushes responsibilities to other governmental actors (e.g., EPA, local governments), and provides only limited resources and support. By not working with local governments early on, Ohio was essentially shifting responsibility to the EPA to enforce the CWA and to municipalities to manage their wastewater systems with limited resources. Further, while Ohio wanted to see more bottom-up watershed management, the lack of funding and support led to implementation barriers. An important takeaway here is that although responsibility to protect the environment can be shifted from states to local governments, this puts local governments in a vulnerable position, where state support is still a prerequisite for success. We should expect this to be representative of the balance of state-local roles when states have neither the political incentives nor administrative capacities to protect the environment.

Making Sense of the State-Local Dimension

In keeping with the previous chapter that examined how different types of states interact with the national government, these four case studies illustrate how different types interact with local governments, which further highlights how states send ripples through environmental federalism. Depending on how states organize their environmental protection efforts, local governments may serve as compliance managers, implementation partners, or be left in a vulnerable position to fill the policy void left as states retreat. Looking at these four case studies highlights two important trends in state-local relations. First, there are many important parallels between issues that occur between national and state governments and between state and local governments, generally to say that state-local relations are as wrought with conflict as national-state relations and how certain states interact with local governments is reminiscent of different eras in national-state relations. This is particularly important to note, as some scholars argue that the state-local dimension is the new battleground of federalism (Riverstone-Newell, 2013; Fowler & Witt, 2019). Although our case studies focus on specific trends that we believe are most prominent, given the diversity of local governments it is likely that they respond to state actions (or inactions) in different ways, leading to a plethora of different types of intergovernmental interactions. It is certainly possible for local environmental agencies within a single state to function as compliance managers, policy innovators, or implementation partners depending on how they interact with the state or the sociopolitical context in which it operates.

Second, local experiences with water quality management and the CWA differ widely, depending on both state actions and the local context, even though they are again facing many of the same legal constraints. In large part, these experiences are shaped by whether political incentives and administrative capacities for environmental protection exist at the state-level, which, in turn, shapes the political, administrative, institutional, and legal context in which local environmental management occurs (see Table 7.1 for a summary). Clearly, in states like California that have both the incentives and capacity to protect the environment at a high level, state institutions so effectively dominate that there

TABLE 7.1 Summary of State Case Studies

Progressive	Struggler
State Domination, Local Compliance Management	*State Collaboration, Local Partners*
California – state dominates water quality regulation through a sophisticated institutional set-up, and local governments are resigned to compliance management	Florida – state supports collaborative watershed management at the regional and community-levels, and local governments function as implementation partners
Delayer	**Regressive**
State Status Quos, Community Political Feasibility	*State Retreat, Local Struggles*
West Virginia – state leverages community-based stakeholder groups to develop politically feasible solutions to local environmental programs that meet national standards	Ohio – state retreats from environmental protection and provides little support to local governments, and local governments are left to figure it out on their own

is little more for local governments to do other than engage in compliance management. On the other hand, in states like Florida, where the incentives exist but not the capacity, states leverage local capacities and treat local governments as implementation partners in order to achieve environmental goals. Conversely, in states like West Virginia, where the capacity exists but not the incentives, states work with community stakeholder groups to find politically feasible solutions to environmental problems. Finally, in states like Ohio, where neither incentives nor capacities exist, states largely retreat from environmental protection, leaving local governments in a vulnerable position to try to comply with federal regulations without state support. Nevertheless, the state-local dimension of environmental federalism is becoming increasingly complex as both governance structures and environmental problems evolve.

References

Agranoff, R. & M. McGuire. 1998. A Jurisdiction-based Model of Intergovernmental Management in US Cities. *Publius* 28(4): 1–20.

Agranoff, R. & M. McGuire. 2001. American Federalism and the Search for Models of Management. *Public Administration Review* 61(6): 671–681.

Bell, S.E. & R. York. 2010. Community Economic Identity: The Coal Industry and Ideology Construction in West Virginia. *Rural Sociology* 75(1): 111–143.

Benham, B., R. Zeckoski, & G. Yagow. 2008. Lessons Learned from TMDL Implementation Case Studies. *Water Practice* 2(1): 1–13.

Borisova, T., L. Racevskis, & J. Kipp. 2012. Stakeholder Analysis of a Collaborative Watershed Management Process: A Florida Case Study. *Journal of the American Water Resources Association* 48(2): 277–296.

Bowen, W.M. 2014. *The Road Through the Rust Belt: From Preeminence to Decline to Prosperity.* Kalamazoo, MI: W.E. Upjohn Institute for Employment Research.

Bowman, A.O'M. & R.C. Kearney. 2011. Second-Order Devolution: Data and Doubt. *Publius* 41(4): 563–585.

Breetz, H.L., K. Fisher-Vanden, L. Garzon, H. Jacobs, K. Kroetz, & R. Terry. 2004. *Water Quality Trading and Offset Initiatives in the US: A Comprehensive Survey.* Dartmouth College and the Rockefeller Center for the US Environmental Protection Agency.

Brody, S.D., W. Highfield, & V. Carrasco. 2004. Measuring the Collective Planning Capabilities of Local Jurisdictions to Manage Ecological Systems in Southern Florida. *Landscape & Urban Planning* 69(1): 33–50.

Byrne, J., K. Hughes, W. Rickerson, & L. Kurdgelashvili. 2007. American Policy Conflict in the Greenhouse: Divergent Trends in Federal, Regional, State, and Local Green Energy and Climate Change Policy. *Energy Policy* 35(9): 4555–4573.

California Environmental Protection Agency. 2019. *About Us* [online]. Available at https://calepa.ca.gov/about/

California Water Boards. 2013. *Fact Sheet* [online]. Available at www.waterboards.ca.gov/publications_forms/publications/factsheets/docs/region_brds.pdf

California Water Boards. 2019. *About the California Water Boards* [online]. Available at www.waterboards.ca.gov/publications_forms/publications/factsheets/docs/board overview.pdf

Chen, Y.C. & K. Thurmaier. 2009. Interlocal Agreements as Collaborations: An Empirical Investigation of Impetuses, Norms, and Success. *American Review of Public Administration* 39(5): 536–552.

Cline, S.A. & A.R. Collins. 2003. Watershed Associations in West Virginia: Their Impact on Environmental Protection. *Journal of Environmental Management* 67(4): 373–383.

Conlan, T. 2006. From Cooperative to Opportunistic Federalism: Reflections on the Half-Century Anniversary of the Commission on Intergovernmental Relations. *Public Administration Review* 66(5): 663–676.

Davis, C. 2014. Substate Federalism and Fracking Policies: Does State Regulatory Authority Trump Local Land Use Autonomy? *Environmental Science & Technology* 48(15): 8397–8403.

Davis, C. & K. Hoffer. 2010. Energy Development in the US Rockies: A Role for Counties. *Publius* 40(2): 296–311.

Dedekorkut, A. 2005. Suwannee River Partnership: Representation Instead of Regulation. In *Adaptive Governance & Water Conflict: New Institutions for Collaborative Planning*, edited by J.T. Scholz & B. Stiftel (pgs. 25–39). Washington, DC: Resources for the Future.

Feiock, R.C. & J.T. Scholz. 2009. Self-organizing Governance of Institutional Collective Action Dilemmas: An Overview. In *Self-organizing Federalism: Collective Mechanisms to Mitigate Institutional Collective Action Dilemmas*, edited by R.C. Feiock & J.T. Scholtz (pgs. 3–32). New York: Cambridge University Press.

Florida Department of Environmental Protection. 2019. *Districts* [online]. Available at https://floridadep.gov/districts

Florida League of Cities. 2011. *Understanding Florida's Home Rule Power* [online]. Available at www.floridaleagueofcities.com/docs/default-source/Civic-Education/historyofhomerule.pdf?sfvrsn=2

Fowler, L. 2016. Local Governments: The "Hidden Partners" of Air Quality Management. *State & Local Government Review* 48(3): 175–188.

Fowler, L. 2018. When Need Meets Opportunity: Expanding Local Air Networks. *American Review of Public Administration* 48(3): 219–231.

Fowler, L. 2019. Is Partnership Quality or Quantity More Effective. *Public Performance & Management Review* 42(5): 1186–1210.

Fowler, L. 2020. Governance, Federalism, and Organizing Institutions to Manage Complex Problems. *Public Administration* [advanced online publication].

Fowler, L. & B. Jones. 2019. Second-order Devolution or Local Activism? Local Air Agencies Revisited. *Review of Policy Research* 36(6): 757–780.

Fowler, L. & G. Rabinowitz. 2019. Balancing Multi-level Politics in Local Environmental Policy Choices. *Public Works Management & Policy* [advanced online publication].

Fowler, L. & S. Witt. 2019. State Preemption of Local Authority: Explaining Patterns of State Adoption of Preemption Measures. *Publius* 49(3): 540–559.

Gerlak, A.K. & T. Heikkila. 2007. Collaboration and Institutional Endurance in US Water Policy. *PS: Political Science & Politics* 40(1): 55–60.

Hicks, W.D., C. Weissert, J. Swanson, & J. Bulman-Pozen. 2018. Home Rule Be Damned: Exploring Policy Conflicts between the Statehouse and City Hall. *PS: Political Science & Politics* 51(1): 26–38.

Hoornbeek, J., E. Hansen, E. Ringquist, & R. Carlson. 2008. *Implementing Total Maximum Daily Loads: Understanding and Fostering Successful Results*. Kent, OH: Center for Public Administration & Public Policy, Kent State University.

Hoornbeek, J., E. Hansen, E. Ringquist, & R. Carlson. 2013. Implementing Water Pollution Policy in the United States: Total Maximum Daily Loads and Collaborative Watershed Management. *Society & Natural Resources* 26(4): 420–436.

Koontz, T.M. & J. Newig. 2014. From Planning to Implementation: Top-Down and Bottom-up Approaches for Collaborative Watershed Management. *Policy Studies Journal* 42(3): 416–442.

Krause, R.M. 2011. Policy Innovation, Intergovernmental Relations, and the Adoption of Climate Protection Initiatives by U.S. Cities. *Journal of Urban Affairs* 33(1): 45–60.

Krause, R.M. 2013. The Motivations behind Municipal Climate Engagement: An Empirical Assessment of How Local Objectives Shape the Production of a Public Good. *Cityscape* 15(1): 125–141.

Krueger, S. & E.M. Bernick. 2010. State Rules and Local Governance Choices. *Publius* 40(4): 697–718.

Langan, T.J. & C.K. McFarland. 2017. City Leadership, City Constraints. *State & Local Government Review* 49(4): 267–274.

League of California Cities. 2019. *What Cities Regulate and How They Regulate* [online]. Available at www.cacities.org/UploadedFiles/LeagueInternet/df/df0d301a-89ee-4cfe-be53-3e06b5985199.pdf

Lubell, M. 2004. Collaborative Watershed Management: A View from the Grassroots. *Policy Studies Journal* 32(3): 341–361.

Lubell, M. & A. Fulton. 2008. Local Policy Networks and Agricultural Watershed Management. *Journal of Public Administration Research & Theory* 18(4): 673–696.

Lubell, M. & L. Lippert. 2011. Integrated Regional Water Management: A Study of Collaboration or Water-Politics-As-Usual in California, USA. *International Review of Administrative Sciences* 77(1): 76–100.

Maddock, T.A. 2004. Fragmenting Regimes: How Water Quality Regulation Is Changing Political-Economic Landscapes. *Geoforum* 35(2): 217–230.

McGuire, M. 2006. Intergovernmental Management: A View from the Bottom. *Public Administration Review* 66(5): 677–679.

Moore, E.A. & T.M. Koontz. 2003. Research Note a Typology of Collaborative Watershed Groups: Citizen-Based, Agency-Based, and Mixed Partnerships. *Society & Natural Resources* 16(5): 451–460.

Nickles. 2013. *Layperson's Guide to California Wastewater*. Sacramento, CA: Water Education Foundation.

O'Leary, R. 1990. The Courts and the EPA: The Amazing "Flannery Decision". *Natural Resources & Environment* 5(1): 18–22.

O'Leary, R. & T. Yandle. 2000. Environmental Management at the Millennium: The Use of Environmental Dispute Resolution by State Governments. *Journal of Public Administration Research & Theory* 10(1): 137–155.

Pautz, M.C. 2010. Front-line Regulators and Their Approach to Environmental Regulation in Southwest Ohio. *Review of Policy Research* 27(6): 761–780.

Percival, R.V. 1987. The Bounds of Consent: Consent Decrees, Settlements and Federal Environmental Policy Making. *University of Chicago Legal Forum* 1987: 327–353.

Potoski, M. & N.D. Woods. 2002. Dimensions of State Environmental Policies: Air Pollution Regulation in the United States. *Policy Studies Journal* 30(2): 208–226.

Reed, D.S. 2014. *Building the Federal Schoolhouse: Localism and the American Education State*. New York: Oxford University Press.

Richardson, J.J. 2011. Dillon's Rule Is from Mars, Home Rule Is from Venus: Local Government Autonomy and the Rules of Statutory Construction. *Publius* 41(4): 662–685.

Riverstone-Newell, L. 2012. Bottom-up Activism: A Local Political Strategy for Higher Policy Change. *Publius* 42(3): 401–421.

Riverstone-Newell, L. 2013. *Renegade Cities, Public Policy, and the Dilemmas of Federalism*. Boulder, CO: First Forum Press.

Ruhl, J.B. 1999. The (Political) Science of Watershed Management in the Ecosystem Age. *Journal of the American Water Resources Association* 35(3): 519–526.

Scheberle, D. 2004. *Federalism and Environmental Policy: Trust and the Politics of Implementation*. Washington, DC: Georgetown University Press.

Shipan, C.R. & C. Volden. 2006. Bottom-up Federalism: The Diffusion of Antismoking Policies from US Cities to States. *American Journal of Political Science* 50(4): 825–843.

Shipan, C.R. & C. Volden. 2008. The Mechanism of Policy Diffusion. *American Journal of Political Science* 52(4): 840–857.

Terman, J.N. & R.C. Feiock. 2015. Improving Outcomes in Fiscal Federalism: Local Political Leadership and Administrative Capacity. *Journal of Public Administration Research & Theory* 25(4): 1059–1080.

U.S. Census. 2019. *Statistical Abstract Series* [online]. Available at www.census.gov/library/publications/time-series/statistical_abstracts.html [Retrieved January 1, 2019].

U.S. Environmental Protection Agency. 2019a. *City of Ironton Clean Water Act Settlement* [online]. Available at www.epa.gov/enforcement/city-ironton-clean-water-act-settlement

U.S. Environmental Protection Agency. 2019b. *City of Youngstown, Ohio, Sewer Overflows Settlement* [online]. Available at www.epa.gov/enforcement/city-youngstown-ohio-sewer-overflows-settlement

U.S. Environmental Protection Agency. 2019c. *Civil Cases and Settlements* [online]. Available at https://cfpub.epa.gov/enforcement/cases/

U.S. Environmental Protection Agency. 2019d. *West Virginia State Review Framework* [online]. Available at www.epa.gov/compliance/west-virginia-state-review-framework

Volden, C. 2005. Intergovernmental Political Competition in American Federalism. *American Journal of Political Science* 49(2): 327–342.

Webb, A.J. 2012. *Municipal Rights: Home Rule in West Virginia* [online]. Available at https://papers.ssrn.com/sol3/papers.cfm?abstract_id=2169493

West Virginia Municipal League. 2019. *Home Rule* [online]. Available at www.wvml.org/programs/home-rule.html

White, S.S. & M.R. Boswell. 2006. Planning for Water Quality: Implementation of the NPDES Phase II Stormwater Program in California and Kansas. *Journal of Environmental Planning & Management* 49(1): 141–160.

Wolman, H., R. McManmon, M. Bell, & D. Brunori. 2008. Comparing Local Government Autonomy Across States. *Proceedings of the Annual Conference on Taxation & Minutes of the Annual Meeting of the National Tax Association* 101: 377–383.

Wood, B.D. & J. Bohte. 2004. Political Transaction Costs and the Politics of Administrative Design. *Journal of Politics* 66(1): 176–202.

Woods, N.D. & M. Potoski. 2010. Environmental Federalism Revisited: Second-order Devolution in Air Quality Regulation. *Review of Policy Research* 27(6): 721–739.

Yates, G.E., T.V. Stein, & M.S. Wyman. 2010. Factors for Collaboration in Florida's Tourism Resources: Shifting Gears from Participatory Planning to Community-based Management. *Landscape & Urban Planning* 97(4): 213–220.

8

SAME STORY, DIFFERENT PROBLEM

In previous chapters, we examined how states have managed legacy environmental programs using a basic assumption: states want to provide environmental protection at the highest level possible within practical limitations. By doing so, we have shown that when it comes to the CAA, CWA, and RCRA, states fall into four categories based on what they determine is the highest level possible (i.e., political incentives) and their practical limitations in doing so (i.e., administrative capacities). In turn, the categories into which states fall help explain both patterns of pollution and intergovernmental conflict stemming from these legacy programs. Now, let us return to the more pressing question with which we began this volume: can any of these lessons be extrapolated to understanding how climate change is or will be managed within the federal system? In other words, does this typology also explain patterns related to climate policy? Before we begin, we should offer a word of caution about the inherent limitations that we face in making inferences about climate change from our existing data. First, our data is based on a specific time period: 2000 to 2010. While the CAA, CWA, and RCRA remained mostly stable during that time and throughout the 2000s, climate policy has dramatically evolved. As such, we must be careful to concentrate our analysis here to that time period while also considering how this may help us explain issues that emerged in the ensuing years or may occur in the future.

Second, as we constructed both the political incentives (Chapter 3) and the administrative capacity (Chapter 4) indices, our focus was on traditional environmental programs, not on climate policies. As such, we must be circumspect in assuming that the same measures that construct political incentives or administrative capacities in that context also explain climate policy.

For instance, our managing information and creating accountability indices are based on program performance from existing environmental programs, which assumes that these same capacities also shape how states administer or will administer climate policies. Additionally, our comparative policy index incorporates toxic releases, but toxic releases likely have little to do with the comparative competitive advantage to which states respond for climate change; thus, we must assume an association between these factors. Consequently, there is an inherent limitation in applying our state categorizations to climate policies. Nevertheless, for the time being and in the interest of simplicity, we will assume that our classification is still reliable within this new context and apply it as such. As we shall see next, our findings related to climate policy are still robust and offer notable insights, despite these constraints.

Explaining Patterns of GHG Emissions

Let us start by considering the most important part of this: does our typology explain patterns of GHG emissions? In order to answer this question, we reconstructed Model 5.4, presented in Chapter 5,[1] using three new dependent variables: 1) concentration of GHG emissions, excluding land use change and forestry (LUCF) (Model 8.1); 2) concentration of GHG emissions, including LUCF (Model 8.2); and 3) concentration of CO_2 emissions (Model 8.3).[2] Note that there are several GHGs tracked that are directly connected to climate change, including CH_4, N_2O, and hydrofluorocarbon gases.[3] CO_2 is the most common, representing more than 85% of GHG emissions from 1990 to 2017. Additionally, LUCF accounts for carbon sinks (i.e., a natural reservoir that stores carbon chemical compounds) associated with forest land, grassland, cropland, or biomass burning, which represent the majority of carbon sinks on a global scale (World Resources Institute (WRI), 2019). In essence, our dependent variables look at this from three perspectives: total GHG emissions produced by anthropogenic sources (i.e., human activity), actual GHG emissions released into the atmosphere (i.e., not offset by natural processes), and the single most prevalent GHG.

Table 8.1 presents results from three regression models. We use the same approach as Model 5.4 (see Table 5.5) by including a group of control variables for industry, state and local spending, population density, wealth, and energy use, as well as correcting for serial correlation issues with the Prais-Winsten estimation (Beck & Katz, 1995). Again, the goal of doing so is to control for any white noise that may obstruct our understanding of the relationship between types of states and GHG emission concentrations. Across all three models, we see that progressive, struggler, and delayer states have lower GHG emissions concentrations than regressive states, and progressive states have lower concentrations than struggler and delayer states. Specifically, coefficients indicate that GHG emissions per capita in progressive states are between 49% and 55% lower

TABLE 8.1 Regression of GHG Emissions[4]

	Model 8.1***	Model 8.2***	Model 8.3***
Progressives	−104.64***	−74.48***	−85.66***
Strugglers	−54.71**	−62.37**	−48.02**
Delayers	−56.38**	−59.29**	−50.03*
Industry	12.29***	12.74***	11.86***
State and Local Spending	−14.29***	−15.01***	−14.38***
Population Density	1.52***	1.47***	1.45***
Wealth	−.003**	−.003*	−.003**
Energy Use	12.97***	13.64***	12.43***
Constant	188.66	150.75	163.76
R-squared	.87	.83	.87
Adjusted R-squared	.87	.83	.87
Durbin-Watson (trans.)	1.30	1.40	1.32
N-size	450	450	450

Note: Statistical significance levels are indicated by *<.05, **<.01, and ***<.001.

than in regressive states, between 14% and 37% lower than in struggler states, and between 17% and 36% lower than in delayer states, *ceteris paribus*. Additionally, in comparison to regressive states, emission concentrations are between 29% and 41% lower in struggler states and between 30% and 39% lower in delayer states. Interestingly, it appears that there is only a marginal difference (i.e., 1% to 3%) between struggler and delayer states; however, further examination indicates it is not a statistically significant difference.

In general, these findings align with both of our findings from Chapter 5 in regard to pollution concentrations and our expectations for how state types compare in terms of emissions concentrations. In other words, the relationship between state types and GHG emissions concentrations is similar to the relationship with pollution concentrations. However, with GHG emissions, this relationship exists without the rigid framework created by national environmental programs, such as the CAA, CWA, or RCRA. Thus, differences in GHG emissions should largely be the product of state-led policy efforts, rather than successful implementation or management of national programs by states. This is an important distinction, because if these differences are already apparent without the structure of a national program, we should expect them to continue as a result of implementing a national program. That is, if progressive states (or delayer/struggler states) are already outpacing regressive states without the resources, structures, or institutionalization that is inherent in existing national environmental programs, we should expect these differences to only persist with the creation of a national climate policy. Of course, we assume that regressive states will not attempt to diligently implement a national climate policy, or at least not to the extent that progressive states will.

Explaining Patterns of State Climate Policy

Of course, we assume that the findings presented earlier are at least partially explained by state-led policy efforts. Thus, let us now consider whether our typology extends to explaining state climate change-related policies adopted during the 2000s.[5] Table 8.2 looks at the distribution of five of the most prevalent state climate policies from the 2000s (Rabe, 2008; Carley, 2011; Carley & Browne, 2013): GHG emissions targets, climate action plans, renewable portfolio standards (RPS), low carbon or alternative fuel standards, and public benefit funds (PBF).[6] Map 8.1 provides a distribution of policy counts across states. First, GHG emissions targets are policies that established a set reduction goal in GHG levels within a specific time period, while climate action plans serve as strategic plans for how to develop and identify cost-effective opportunities for reducing GHG emissions. Notably, emissions targets and climate action plans

TABLE 8.2 Count of Climate Policies by State Type

	Progressives (15 States)	Stragglers (12 States)	Delayers (11 States)	Regressives (12 States)	Total
GHG Emissions Targets	7 (46.7%)	3 (25.0%)	1 (9.1%)	1 (8.3%)	12
Climate Action Plans	8 (53.3%)	9 (75.0%)	5 (45.5%)	2 (16.7%)	24
Renewable Portfolio Standards	11 (73.3%)	8 (66.7%)	7 (63.6%)	6 (50.0%)	32
Low Carbon/Alternative Fuel Standards	3 (20.0%)	1 (8.3%)	1 (9.1%)	1 (8.3%)	6
Public Benefits Fund	8 (53.3%)	6 (50.0%)	2 (18.2%)	4 (33.3%)	20
Average Policies per State	2.5	2.3	1.5	1.2	

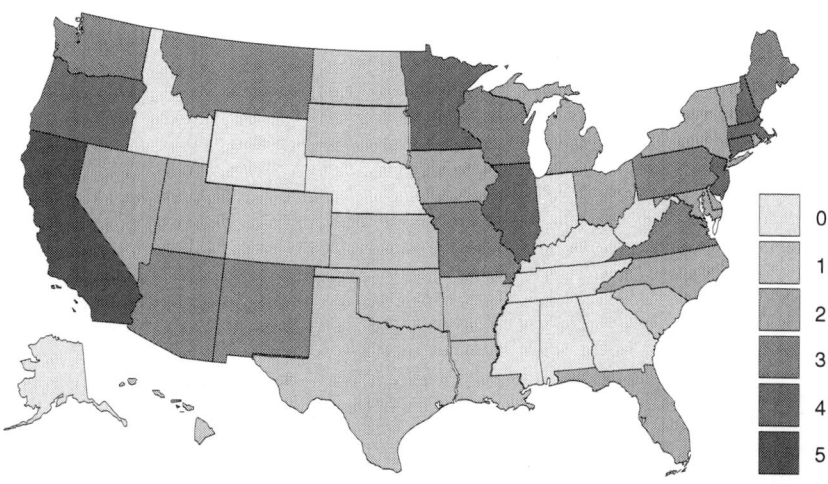

MAP 8.1 Number of Climate Policies by State

are the state policies most directly focused on mitigating the sources of climate change; while other policies on this list are considered climate policies and make a considerable impact on reducing GHG emissions, they are not as direct in doing so as emissions targets or action plans. In both cases, policies are concentrated in progressive and struggler states, as opposed to delayer or regressive states. Interestingly, climate action plans are the only policy category that progressive states do not dominate in terms of either total states or portion of states.

Second, RPS and low carbon or alternative fuel standards are both regulatory policies aimed at encouraging a shift in fuel sources away from GHG-producing fossil fuels. The former is specifically geared towards electricity generation, and the latter towards transportation fuels. RPS was one of the most popular (and examined) innovations in energy policy adopted by states during the late 1990s and 2000s, with a majority of states now operating some form of them. On the other hand, alternative fuel standards were less common during this time frame, although it has caught on in more states since (Carley, 2011; Carley & Browne, 2013). The distribution of policies across states again shows progressive states outpacing all other categories in the number of operations. Interestingly, there appears to be far less separation across the other categories of states. Finally, PBFs provide financial assistance to utilities for the development of energy efficiency or renewable energy initiatives and are funded through a small charge on utility bills. The goal of PBFs is to offset the high capital costs associated with research and development for new technology and system-level technology adoptions. Again, we see a similar pattern of adoptions across states, with progressive states leading the pack. Interestingly, this is the only policy category that regressive states are not the least likely to adopt.

Looking at this from a slightly different perspective, progressive and struggler states average about 2.5 and 2.3 policies adopted per state, as compared to delayer and regressive states, which average about 1.5 and 1.2 policies per states, respectively. Additionally, a majority of progressive states (60%) have adopted three or more of these policies, and a majority of struggler states (91.7%) have adopted two or more, as compared to delayer and regressive states in which a majority of states (63.6% and 66.7%, respectively) have adopted one or none. While there are relatively clear patterns that emerge from looking at these five types of policies, it is important to note that there are other climate-related policies, including a plethora of financial incentives, that are not included here (Byrne, Hughes, Rickerson, & Kurdgelashvili, 2007; Carley, 2011; Carley & Browne, 2013). Also, if we look beyond the 2000s, we find that additional states adopted some of these policies, and there have been other policy innovations as well (e.g., carbon pricing) (CCES, 2019; DSIRE, 2019). As such, this table may not paint a full picture of the complexity of state climate policies, but it does serve its primary goal of highlighting the variation in how states have adopted climate policies during our specific time frame.

Given that, this breakdown would largely suggest that progressive and struggler states were much more proactive in regard to climate policies during the 2000s than delayer and regressive states. In general, this likely offers a partial explanation for the relatively lower concentrations of GHG emissions in progressive states compared to other states. That is, progressive states took the initiative to adopt climate policies, and, as a result, reduced their relative GHG production. Furthermore, this may also provide us with some additional understanding of why there is only a marginal difference between emissions concentration in delayer versus struggler states. Although struggler states may be more willing to adopt policies in response to the inherently pro-environmental political context, delayer states may be able to make just as much of an impact on GHG emissions with fewer policies as a result of their superior administrative capacities. In other words, struggler states are adopting several policies that are only somewhat successful, while delayer states are adopting fewer policies but are experiencing a bigger impact from each. Consequently, this may suggest that delayer states are poised to outperform struggler states under a national climate policy, even if they are unlikely to match progressive states.

Intergovernmental Relations in the Climate Era

Now, let us turn our attention to the intergovernmental component. Rabe (2011) argues that there have essentially been three eras of intergovernmental relations when it comes to climate policy in the US. From 1975 to 1997, neither federal nor state governments were particularly involved in making climate policy, and any attempts were largely symbolic. It was not until around 1998 in which the states began to assert themselves, and largely dominated this policy arena until 2007. State domination was chiefly a result of a lack of federal involvement. In other words, states were left to fill a policy vacuum as lawmakers, experts, and citizens began to take climate change more seriously. Finally, starting around 2008, the national government under the Obama administration began asserting itself and challenged state leadership, leading to an era of contested federalism. Consequently, from the late 1990s to the late 2000s, the story of US climate policy focuses on state innovations and the asymmetries in which states were adopting policies. Much of this early behavior was the result of strategic economic or political advantages in both horizontal and vertical intergovernmental competition. In the late 2000s, the narrative shifted to focus on the conflicts between national and state governments as both jockeyed for who would drive the climate policy bus (Peterson, 2004; Rabe, 2011).

As such, it is during the mid to late 2000s that we start to see the stage set for patterns of intergovernmental conflict over climate policy. For instance, California and Massachusetts (progressive states) both established themselves early on as leaders on climate policy and used this position to challenge the national government's lack of action. Of course, Massachusetts launched a landmark case (Massachusetts v. EPA, discussed in Chapter 6) that signaled a turning point in US climate policy. On the other hand, not only would California's elected officials, such as Governor Arnold Schwarzenegger and Air Resources Board Chair Mary Nichols, launch political attacks on the national government for not acting on climate change, but also the state legislature passed one of the nation's most aggressive climate change laws, and state agencies requested waivers under the CAA to set their own standards for CO_2 emissions and petitioned the EPA to regulate GHG emissions. Undoubtedly, these states were trying to push the national government into acting on climate change but found a less-than-receptive audience under President Bush. Tensions cooled to some extent once President Obama took over the White House, but progressive states did not back down from challenging the national government in the ensuing years (Peterson, 2004; Byrne, Hughes, Rickerson, & Kurdgelashvili, 2007; Rabe, 2007, 2008, 2011; Harrison, 2013).

Struggler states followed a different path and instead sought ways to leverage collaborations to pursue their interests rather than engage in open warfare. For instance, many states found a strategic advantage in regional partnerships, such as the Regional Greenhouse Gas Initiative in the Northeast, the Midwestern Greenhouse Gas Reduction Accord, or the Western Climate Initiative (Rabe, 2008, 2011; Harrison, 2013). Naturally, eight of 12 struggler states joined one of these regional collaborations (compared to ten progressive, two delayer, and three regressive states), and of the remaining four, all were located in the South/Southeast, where a regional partnership did not exist (Conservation in Changing Climate, 2019a, 2019b, 2019c). Other states, like Michigan and Wisconsin, negotiated with the national government for "preferred treatment for themselves under any expanded federal policy" (Rabe, 2011, p. 507). Specifically, Wisconsin lobbied Congress for flexibility in carbon sequestration credits, support for renewable energy research and development projects, and revenue-sharing from cap-and-trade. On the other hand, Michigan received significant support under the American Recovery and Reinvestment Act (ARRA) for clean energy development and for the state's embattled auto industry (Rabe, 2011). In both cases, these struggler states were willing to cede ground to the national government in return for additional support in accomplishing state policy goals.

Regressive states were largely quiet on climate policy until the shift towards more national policy action, when states began to push back against total or partial preemption efforts. Generally, regressive states advocated against federal overreach and for decentralization, which would allow them to evade dealing with the policy problem and shift compliance costs elsewhere. Texas served as the poster child by petitioning the EPA to stop regulating GHG emissions in general, requesting waivers from ethanol regulations, and threatening legal action against further national initiatives (see Chapter 6) (Rabe, 2011). In other words, "Texas [was] clearly drawing a line in the sand" (Pendergrass, 2010).

Although other regressive states were not as proactive, they followed a similar pattern of resistance and conflict when national policies arrived at their doorstep. In contrast, delayer states remained relatively quiet and have largely positioned themselves to avoid conflict (both internally and externally) over climate policy. Other than North Dakota, South Dakota, and Utah siding with the EPA in Massachusetts v. EPA in an effort to uphold the status quo in climate policy, delayer states have not gained much attention and are rarely discussed either anecdotally or as the subject of case studies, which suggests a lack of substantive events originating from these states (e.g., Rabe, 2007, 2008, 2011). In sum, patterns of national-state relations surrounding climate policy are consistent with those surrounding legacy environmental problems.

On the flip side, patterns of state-local interactions may look somewhat different than our initial expectations. Although scholars have become particularly interested in local climate initiatives in recent years, little of this scholarship examines the intergovernmental dimensions that shape local actions, so there are fewer insights here (Schreurs, 2008; Urpelainen, 2009; Krause, 2011, 2013; Sharp, Daley, & Lynch, 2011; Kwon, Jang, & Feiock, 2014; Krause, Yi, & Feiock, 2016). In general, previous scholarship suggests that two trends dictate local government roles in environmental policy. From the top-down perspective, states use local governments to make up for their lack of administrative capacities or to shift political blame for issues with which they are disinterested in dealing. From the bottom-up perspective, local governments respond to policy voids left by a lack of state action and to horizontal competitive pressures (Agranoff & McGuire, 2001; Volden, 2005; Shipan & Volden, 2006, 2008; Feiock & Sholz, 2009; Riverstone-Newell, 2013; Fowler & Jones, 2019). Looking specifically at climate policies, Krause (2011) finds that only the latter explains whether cities are formally committed to GHG reduction policies; specifically, local-level characteristics and horizontal diffusion are the primary predictors.

But this may all be complicated by the rather fluid state that climate policy is in and has been in since the mid-2000s. Krause, Yi, and Feiock (2016) explain it as such:

> For several years, [the local climate policy innovation movement] was accelerating rapidly; the number of municipalities with climate protection commitment increased almost fivefold between 2005 and 2010. Since 2010, however, the momentum on this policy has shifted: fewer new cities are adopting climate protection initiatives and, in noticeable numbers, existing ones are being abandoned.
>
> *(p. 189)*

Local government responses to climate change are at least partially a function of balancing out state actions, so as states have navigated the climate policy odyssey with the national government, local governments have been left in limbo. Consequently, some early adopters of climate policies later terminated those policies due to retrenchment of political support or when facing the administrative realities of effectively running programs (Krause, Yi, & Feiock, 2016). In terms of the former, as states play the blame avoidance-credit claiming game, the political rewards at the local-level shift as well, with community participation, organized political interests, and longevity of policy networks being key ingredients for successful local programs (Volden, 2005; Sharp, Daley, & Lynch, 2011).

Basically, roles of national, state, and local governments in climate policy remain up in the air, with shifts in state policies sending shockwaves through the federal system. As such, it is difficult to determine if local governments are currently or will eventually follow patterns previously observed for legacy environmental programs once stability is achieved at national- or state-levels, or if this is a new pattern of behavior altogether. From the mid-2000s to the early 2010s, local governments from all 50 states adopted some form or another of climate policies, but, in many cases, those policies were largely symbolic, impacts at local- or global-levels are suspect, and many policies have since been terminated (Krause, 2011; Sharp, Daley, & Lynch, 2011; Krause, Yi, & Feiock, 2016). Consequently, it is difficult to draw conclusions about how state-local relations surrounding climate policy compare or contrast to those observed for legacy environmental policies when policies such as the CAA and CWA benefit from decades of institutionalized structures, flushed out roles for national, state, and local governments, and policy learning. Of course, more clarity is likely to be gained over time, but this remains an important open question, as many scholars argue that the state-local dimension is the new battleground of federalism (Riverstone-Newell, 2017; Fowler & Witt, 2019).

Another Brief Thought Experiment

At this point, we believe it is a fruitful exercise to conduct another brief thought experiment in order to determine how this data may help us understand how a national climate policy would operate. Based on our analysis (Table 8.1), we can make some inferences about the potential performance comparisons between different types of states. Specifically, a national climate policy would likely set broad national standards for GHG targets, which would require states to undertake policy and administrative actions in order to reduce their total state-wide emissions over a set period of time. Under this scenario, it is not a logical leap to assume that progressive states are likely to outperform all other types of states in terms of meeting and/or exceeding national standards, and struggler and delayer states will likely do the same in comparison to regressive states. In other words, we can expect progressive states to do a better job of reaching and potentially exceeding GHG emissions reduction targets than other types of states, and regressive states to comparatively flounder. If we use findings from Model 8.2 (examining GHG emissions including LUCF) to extrapolate how performance would likely compare, we could then infer that progressive states are likely to outperform regressive states by roughly 49.4%, delayer states by 16.6%, and struggler states by 13.7%. We could make similar inferences about how struggler, delayer, and regressive states compare to each other.

By extension, we can use these comparisons to determine what our expectations should be for overall GHG emission reductions if states were to follow this pattern while implementing a national policy. Before we can do that, we need to set out a few assumptions. First, let us assume that the overall goal is a 26% reduction below 2005 emissions levels, which is called for by the Paris Agreement. While President Trump withdrew the US from the Paris Agreement in 2019, its principal goals and framework still serve as the international standard for climate action. Notably, 24 states have joined the United States Climate Alliance, which is dedicated to advancing the goals of the Paris Agreement (US Climate Alliance, 2019). Those 24 states include 11 progressive states, eight strugglers, two delayers, and three regressives, which is keeping with the policy adoption patterns examined here. As such, our goal here is to determine what performance expectations would states need to meet and/or what national reduction targets would need to be set in order to realize a 26% reduction.

Second, let us assume that reductions would come in the form of GHG emissions concentrations, which would then be translated into lower overall emissions. Why is this important? As Map 8.2 illustrates, states with large urban populations and industrialized economies naturally produce more GHG emissions than smaller rural states with more agricultural production, especially when LUCF offsets are considered. While it is more accurate for us to understand how states compare to each other via emissions rates (i.e., concentration),

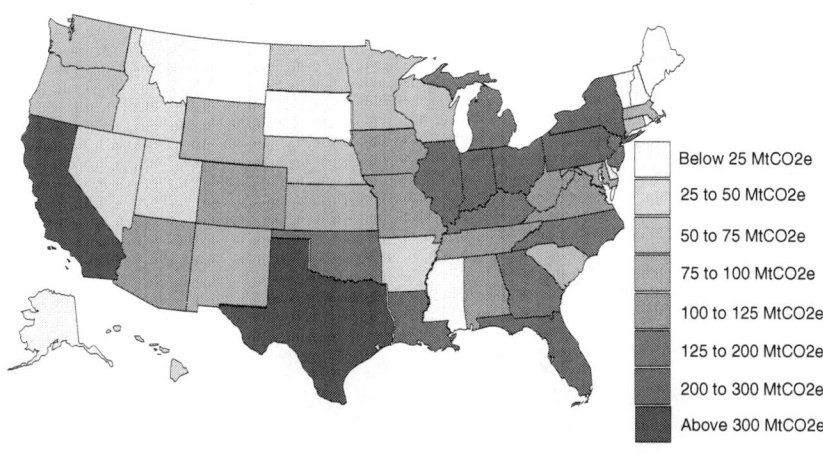

Below 25 MtCO2e

25 to 50 MtCO2e

50 to 75 MtCO2e

75 to 100 MtCO2e

100 to 125 MtCO2e

125 to 200 MtCO2e

200 to 300 MtCO2e

Above 300 MtCO2e

MAP 8.2 Total GHG Emissions (Including LUCF) by State

the goal of GHG targets focuses on total emissions, not concentrations. In other words, a 26% reduction in GHG emissions equals only a 26% total emissions reduction if we assume that there is equal reduction in every state. But if we assume there is an unequal amount of reduction in every state, larger reductions in certain states can balance out smaller reductions in other states. For instance, based on 2005 data, a 50% reduction from the ten highest producing GHG emission states would meet the national goal, even if other states experience no reduction. Consequently, we assume that performance comparisons explain reductions in emission concentrations, and those reductions then manifest in total concentrations. Finally, let us ignore growth and focus only on 2005 data. While economic and population growth are important confounding factors here, it adds unnecessary complications to our thought experiment, so we will set aside those issues for a moment.

Based on these assumptions, we calculated the total national-level reduction in 2005 GHG emissions (including LUCF) based on four scenarios: 26%, conservative, moderate, and idealistic (see Table 8.3). First, the 26% scenario assumes that the national target reduction is set at 26% and the realized target reduction is also 26%. In order to achieve this, progressive states would have to achieve 135% of national standards, struggler states 116.5%, delayer states 112.6%, and regressive states 68.3%. In other words, under this scenario, progressive, struggler, and delayer states all exceed the national standards, while regressive states only achieve about two-thirds of the goal. Second, the conservative scenario assumes that only progressive states actually achieve the set

TABLE 8.3 Scenarios for National Climate Policy Implementation

	26% Scenario	Conservative Scenario	Moderate Scenario	Idealistic Scenario
Required Performance in Comparison to Reduction Target (based on findings from Model 8.2)				
Progressives	135.0%	100%	115.9%	197.7%
Strugglers	116.5%	86.3%	100%	170.6%
Delayers	112.6%	83.4%	96.6%	164.8%
Regressives	68.3%	50.6%	58.6%	100%
Realized Reduction If Target Is Set at 26%	26%	19.3%	22.3%	44.4%
Necessary Target to Realize 26% Reduction	26%	35.1%	30.3%	17.8%
Required Performance in Comparison to Reduction Target (based on findings from Model 5.4)				
Progressives	135.7%	100%	123.6%	178.7%
Strugglers	102.9%	75.8%	93.7%	135.5%
Delayers	109.8%	80.9%	100%	144.7%
Regressives	75.9%	55.9%	69.1%	100%
Realized Reduction If Target Is Set at 26%	26%	19.2%	23.7%	34.3%
Necessary Target to Realize 26% Reduction	26%	35.3%	28.6%	19.8%

national standard, and all other states fail to do so. Under this scenario, if the reduction target remained at 26%, the realized reduction would be only 19.3%, and in order to achieve a realized reduction of 26%, the national goal would have to be 35.1%.

Third, the moderate scenario assumes that delayer states will meet the national standards, and other states perform comparatively. Given our expectations that delayer states will largely comply with national standards, this serves as a good middle-of-the-road scenario. In order to offset underperformance by some states, progressive states would have to achieve 115.9% of national standards, while regressive states would fall significantly short at 58.6%. Under this scenario, if the reduction target remained at 26%, the realized reduction would be only 22.3%, and in order to achieve a realized reduction of 26%, the national goal would have to be 30.3%. Finally, the idealistic scenario assumes that regressive states meet the national standard and all other states drastically exceed it by 64.8% to 97.7%. Under this scenario, if the reduction target remained at 26%, the realized reduction would be 44.4%, and in order to realize a reduction of 26%, the national goal would have to be only 17.8%.

A key problem here is that we have no way of estimating how a national climate policy may affect the comparative performance. So, let us supplement this by taking these same scenarios but substituting the performance comparisons from Model 8.2 with those from Model 5.4. In doing so, this provides us further insights into how this may work under the structure of a national policy. Notably, a key difference is that with externally mandated standards, delayer states should be expected to perform better in comparison to other states. In general, both performance expectations for types of states across scenarios and reduction targets are similar between our two sets of performance expectations. However, two key differences are that expectations for regressive states would be higher under a national climate policy, and on average lower for every other type of state, especially struggler states. Essentially, our expectations for regressive states based on Model 8.2 may be too low, and when we raise those expectations based on Model 5.4, expectations for other states can be relaxed. Only the idealistic scenario involves a major shift (10%+) in expectations across several categories, but that shift largely suggests a national climate policy could be successful if it mimics performance under legacy environmental programs. In other words, regardless of the performance expectations on which we rely, our findings are fairly consistent.

But are any of these scenarios realistic? One could argue that the vast majority of states have met the basic compliance standards set out by the CWA, CAA, and RCRA and many states have gone above and beyond national standards, so the idealistic scenario is not unobtainable. On the other hand, one could also argue that there is evidence of systematic failure in meeting those goals, so the idealistic scenario is a pipe dream. The conservative scenario is likely a worst case one, and we should probably expect more than only 15 states to achieve national targets. Frankly, if our expectations for success are so dismal, it would beg a larger question of whether this is a policy avenue worth pursuing in general. The moderate, or 26%, scenarios are likely the most realistic in terms of expectations for both state-level performance and national goal targets. Under any scenario though, we can likely make two inferences: 1) progressive states will have to overachieve in order to balance out other states, namely regressive states; and 2) national target goals will have to shift in response to our performance expectations for states. In other words, regardless of which scenario we prefer, we should expect unequal performance across states, and national targets will have to be constructed accordingly.

But there are some important flaws here. First, and most importantly, we do not consider the role of economic growth, how the climate change picture has evolved since 2005, or the timeline in which these reductions can or should be achieved. Second, we have no way of knowing how a national climate policy may affect how states operate in this policy space or how intergovernmental relations may evolve as a result. For instance, it may give the administrative efforts of delayer states more backbone; or, struggler states may find implementation partners at the local-level to balance their administrative capacities and boost their efficacy; or, we may find that intergovernmental conflicts create gridlock throughout the whole system as state and national governments jockey for control. Third, we may also find that the administrative and political realities surrounding climate change have shifted significantly since 2005, so states may not fit as readily into our categories. Of course, this is a highly speculative exercise, so the results should be taken with caution. The real value of this thought experiment though is simply understanding that if we expect patterns from legacy environmental policies to be consistent with a national climate policy, we must understand that there will also be large asymmetries in how states implement it, and that will dictate how successful it is.

Looking Forward to a National Climate Policy

So, let us now try to answer the question with which we started this chapter: can any of the lessons learned from managing legacy environmental problems be extrapolated to understanding how climate change is or will be managed within the federal system? The short answer is yes. Looking at GHG emissions, climate policy adoptions, and intergovernmental relations, we can largely conclude that they follow a similar pattern to those observed for the CAA, CWA, and RCRA examined in previous chapters. Therefore, we can continue to expect states to be the linchpin in the system, with some being more effective and others being less so depending on the combination of political incentives and administrative capacities that shape the context in which environmental protection is delivered. Most obviously, this means that progressive states are going to be policy innovators that challenge the national government to do more, while regressive states are going to drag their heels and try to stop federal intrusion into their territory; delayer and struggler states will be caught somewhere in the middle as they try to avoid conflicts over climate policy and work the system to their advantage. The two caveats of this conclusion are that we are limiting our observations to the 2000s and our state classifications to measures developed in Chapters 3 and 4. However, we wholly expect these patterns to have continued throughout the 2010s, even if the political or administrative contexts of particular states have evolved.

Finally, let us also consider an equally important question: how successful can we expect a national climate policy to be? The short answer is: it probably will not lead to worse outcomes. We base that conclusion on two points. First, our experience with legacy environmental problems indicates that establishing national standards has generally improved environmental protection, even if they are failures in particular jurisdictions. Or, in Weibust's (2016) words:

> Decentralized environmental governance, without federal floor standards, was not effective in the United States. Air and water quality in the US deteriorated steadily and this declined was not halted or reversed until federal legislation set minimum national standards in the 1960s.
>
> *(p. 191)*

Second, we are cautious about how successful a national climate policy may be, because it will largely depend on state implementation. Despite suggestions that a national climate policy can be managed through a top-down approach, the federal system is not top-down oriented now, and the national government is wholly reliant on states to implement environmental policies. As such, if progressive states are aggressive enough in their actions and take on enough of the burden to offset underachieving states, we should expect success; if not, it is highly unlikely that national goals will be met. In sum, similar to our legacy programs, states will likely remain the linchpins in the federal system, and how they respond to a national policy will dictate whether goals are obtained.

Notes

1. I use five variables to account for socioeconomic factors that affect GHG emissions. Three are measured in the same way as in Chapter 5: 1) industry per capita from the manufacturing, mining, utilities, and waste management sector in 2009 dollars; 2) state and local total per capita expenditures in 2009 dollars; and 3) per capita personal income in thousands of 2009 dollars. Again, I used data from Census (2019) and BEA (2019). While I use population as a control factor in Chapter 5, I alter this slightly and use population density here instead. Finally, using data from WRI (2019), I also control for energy use per capita measured as tons of oil equivalent. While there are other socioeconomic, political, and technical variables that other scholars also use as control variables, I focus on these five in order to maintain consistency with my analysis from Chapter 5 and to prioritize parsimony in order to keep the focus on my typology.

2. I initially measured emissions concentration as toxic metric tons of carbon dioxide equivalent (MTCO2e) per square mile, but I later converted metric tons to 1.0×10^{-5} metric tons by dividing the originally reported MTCO2e by 100,000. MTCO2e is the most common measure of GHG emissions or other climate change emissions indicators, and I converted it to aid in interpretation, as initial data produced coefficients below .0001. I obtained data from the WRI (2019), which compiled the dataset based on data from the US Census Bureau, US Department of Commerce, and EPA.

3. Commonly referred to as F-gases, these include, but are not limited to, tetrafluoromethane (CF_4), hexafluoroethane (C_2F_6), sulfur hexafluoride (SF_6), and nitrogen trifluoride (NF_3).

4. I use OLS regression. Similar to Models 5.3 and 5.4 from Chapter 5, diagnostic tests indicated a potential issue with serial correlation, which I corrected for using the Prais-Winsten estimation. I report the transformed Durbin-Watson statistic, which falls into acceptable critical values for both models (Beck & Katz, 1995).

5. For the purposes of this discussion, I also include policies that were adopted during the late 1990s, as states would be unlikely to then readopt a policy in the 2000s; however, I do not include any policy adopted after 2010.

6. I obtained data on specific state policies from the Center for Climate and Energy Solutions (CCES) (2019), and North Carolina Clean Energy Technology Center's Database of State Incentives for Renewables and Efficiency (DSIRE) (2019).

References

Agranoff, R. & M. McGuire. 2001. American Federalism and the Search for Models of Management. *Public Administration Review* 61(6): 671–681.

Beck, N. & J.N. Katz. 1995. What to Do (and Not to Do) with Time-Series Cross-Section Data. *American Political Science Review* 89(3): 634–647.

Byrne, J., K. Hughes, W. Rickerson, & L. Kurdgelashvili. 2007. American Policy Conflict in the Greenhouse: Divergent Trends in Federal, Regional, State, and Local Green Energy and Climate Change Policy. *Energy Policy* 35(9): 4555–4573.

Carley, S. 2011. The Era of State Energy Policy Innovation: A Review of Policy Instruments. *Review of Policy Research* 28(3): 265–294.

Carley, S. & T.R. Browne. 2013. Innovative US Energy Policy: A Review of States' Policy Experiences. *WIREs Energy & Environment* 2(5): 488–506.

Center for Climate & Energy Solutions. 2019. *State Climate Policy Maps* [online]. Available at www.c2es.org/content/state-climate-policy/ [Retrieved January 1, 2019].

Conserving a Changing Climate. 2019a. *Midwestern Greenhouse Gas Reduction Accord* [online]. Available at https://climatechange.lta.org/midwestern-accord/ [Retrieved January 1, 2019].

Conserving a Changing Climate. 2019b. *Regional Greenhouse Gas Initiative* [online]. Available at https://climatechange.lta.org/regional-greenhouse-gas-initiative/ [Retrieved January 1, 2019].

Conserving a Changing Climate. 2019c. *Western Climate Initiative* [online]. Available at https://climatechange.lta.org/western-climate-initiative/ [Retrieved January 1, 2019].

Database for State Incentives for Renewable & Efficiency (DSIRE). 2019. *Programs. North Carolina Clean Energy Technology Center* [online]. Available at https://programs.dsireusa.org/system/program

Feiock, R.C. & J.T. Scholz. 2009. Self-Organizing Governance of Institutional Collective Action Dilemmas: An Overview. In *Self-Organizing Federalism: Collective*

Mechanisms to Mitigate Institutional Collective Action Dilemmas, edited by R.C. Feiock & J.T. Scholz (pgs. 3–32). New York: Cambridge University Press.

Fowler, L. & B. Jones. 2019. Second-Order Devolution or Local Activism? Local Air Agencies Revisited. *Review of Policy Research* 36(6): 757–780.

Fowler, L. & S. Witt. 2019. State Preemption of Local Authority: Explaining Patterns of State Adoption of Preemption Measures. *Publius* 49(3): 540–559.

Harrison, K. 2013. Federalism and Climate Policy Innovation: A Critical Reassessment. *Canadian Public Policy* 39(2): 95–108.

Krause, R.M. 2011. Policy Innovation, Intergovernmental Relations, and the Adoption of Climate Protection Initiatives by U.S. Cities. *Journal of Urban Affairs* 33(1): 45–60.

Krause, R.M. 2013. The Motivations behind Municipal Climate Engagement: An Empirical Assessment of How Local Objectives Shape the Production of a Public Good. *Cityscape* 15(1): 125–141.

Krause, R.M., H. Yi, & R.C. Feiock. 2016. Applying Policy Termination Theory to the Abandonment of Climate Protection Initiatives by U.S. Local Governments. *Policy Studies Journal* 44(2): 176–195.

Kwon, M., H.S. Jang, & R.C. Feiock. 2014. Climate Protection and Energy Sustainability Policy in California Cities: What Have We Learned? *Journal of Urban Affairs* 36(5): 905–924.

North Carolina Clean Energy Technology Center. 2019. *Database of State Incentives for Renewables and Efficiency* [online]. Available at www.dsireusa.org/ [Retrieved January 1, 2019].

Pendergrass, J. 2010. Texas Tailors the Tailoring Rule. *Environmental Forum* 27(5): 10.

Peterson, T.D. 2004. The Evolution of State Climate Change Policy in the United States; Lessons Learned and New Directions. *Widener Law Journal* 14: 81–120.

Rabe, B. 2007. Environmental Policy and the Bush Era: The Collison between the Administrative Presidency and State Experimentation. *Publius* 37(3): 413–431.

Rabe, B. 2008. States on Steroids: The Intergovernmental Odyssey of American Climate Policy. *Review of Policy Research* 25(2): 105–128.

Rabe, B. 2011. Contested Federalism and American Climate Policy. *Publius* 41(3): 494–521.

Riverstone-Newell, L. 2013. *Renegade Cities, Public Policy, and the Dilemmas of Federalism*. Boulder, CO: First Forum Press.

Riverstone-Newell, L. 2017. The Rise of State Preemption Laws in Response to Local Policy Innovation. *Publius* 47(3): 403–425.

Schreurs, M.A. 2008. From the Bottom Up: Local and Subnational Climate Change Politics. *Journal of Environment & Development* 17(4): 343–355.

Sharp, E.B., D.M. Daley, & M.S. Lynch. 2011. Understanding Local Adoption and Implementation of Climate Change Mitigation Policy. *Urban Affairs Review* 47(3): 433–457.

Shipan, C.R. & C. Volden. 2006. Bottom-Up Federalism: The Diffusion of Anti-smoking Policies from U.S. Cities to States. *American Journal of Political Science* 50(4): 825–843.

Shipan, C.R. & C. Volden. 2008. The Mechanisms of Policy Diffusion. *American Journal of Political Science* 52(4): 840–857.

Urpelainen, J. 2009. Explaining the Schwarzenegger Phenomenon: Local Frontrunners. *Global Environmental Politics* 9(3): 82–105.

U.S. Bureau of Economic Analysis. 2019. *Regional Data* [online]. Available at https://apps.bea.gov/itable/iTable.cfm?ReqID=70&step=1 [Retrieved January 1, 2019].

U.S. Census. 2019. *Statistical Abstract Series* [online]. Available at www.census.gov/library/publications/time-series/statistical_abstracts.html [Retrieved January 1, 2019].

U.S. Climate Alliance. 2019. *About Us* [online]. Available at www.usclimatealliance.org/about-us [Retrieved January 1, 2019].

Volden, C. 2005. Intergovernmental Political Competition in American Federalism. *American Journal of Political Science* 49(2): 327–342.

Weibust, I. 2016. *Green Leviathan: The Case for a Federal Role in Environmental Policy.* New York: Routledge.

World Resources Institute. 2019. *US States Greenhouse Gas Emissions* [online]. Available at http://datasets.wri.org/dataset/cait-us-states-greenhouse-gas-emissions [Retrieved January 1, 2019].

9

LESSONS LEARNED

In the preceding chapters, we have taken an in-depth look at: the history of environmental federalism and why states play such important roles (Chapter 2); the political reasons why states protect the environment (Chapter 3) and the administrative realities of doing so (Chapter 4); how the political and administrative dimensions shape state behaviors surrounding legacy environmental programs (Chapters 5, 6, and 7); and whether we can expect these behaviors to extend to climate policy (Chapter 8). In general, our examination indicates that not all states take the same approach to environmental protection, which creates asymmetries across the nation. In turn, these asymmetries have important implications for how successful federal environmental programs (i.e., CAA, CWA, RCRA) have been, and how successful we should expect a national climate policy to be. Suffice it to say that some states are effective in protecting the environment as well as responding to new challenges as they emerge, and some states are largely disinterested in pursuing environmental protection as a policy goal. Still, many states are stuck somewhere in the middle, as political incentives encourage delayed action or administrative limitations lead to struggles and represent an important constituency in determining whether national policies are a systematic success or failure.

From Old Legacies to New Challenges

The concept of federalism has been ingrained into the fabric of American government since the Founding Fathers organized a novel system that allowed for sovereign power to be shared by both national and state governments. While we have evolved from rural, agrarian-oriented communities into a post-industrial, urbanized economic state over the last two centuries, this institutional approach still forms the framework of how our society is governed. Over the ensuing decades, new policy problems have challenged the utility of this framework, and power has shifted back and forth from national- to state-centric as both levels of government have jockeyed for control over regulatory authorities. It was during the mid-20th century that the federal framework was applied to the emerging environmental challenges that came along with industrial and urban growth, forming the basis for modern environmental federalism. Although it may not be the ideal system for managing environmental problems, federalism is the institutional framework by which government in the US operates, and has positioned states as the linchpins that play an ever-important role in connecting the national to the local. Even though it may not always be the most administratively efficient solution to solving wicked environmental problems, it does serve to ameliorate political conflicts that stretch back to the founding of America.

While environmental problems refuse to adhere to the geographic boundaries drawn between states, over the decades federalism has adapted to create an efficacious regulatory regime for legacy environmental problems. Notwithstanding the criticisms that abound, research indicates that air, land, and water quality have at best improved significantly since the 1970s and at worst have remained stable, despite the fact that economic development has put a significant strain on our natural resources during that time. Subsequently, many would argue that programs like the CAA and CWA have been successful in maintaining and/or improving environmental quality across the American continent (Fowler, 2014). Consequently, it is likely that we will continue to utilize this institutional approach to address the next great environmental challenge: climate change. Despite the various proposals that have been thrown around for a national climate policy, there are no concrete answers yet for what form one may take. But if we look at proposals like the Clean Energy Plan, the American Recovery and Reinvestment Act, or even the Green New Deal, we can infer that national climate policy would follow the blueprint of existing environmental and economic programs in which the national government sets guidelines and provides resources to states, which then develop implementation plans and put programs into practice (Konisky & Woods, 2016, 2018).

Our typology generally suggests that under this system, state behavior is dictated by two complementary dimensions: political incentives and administrative capacities. On the one hand, how much environmental protection is needed or wanted is chiefly a political question. It is driven by how policy actors see the environment in terms of problems (i.e., social construction) and the responsibilities of government in addressing those problems. To this end, political incentives shape how state leaders determine what is the highest level of environmental protection that should be offered. In other words, political incentives push state officials to make a commitment to protecting the environment, but in the absence of incentives, the same officials are likely to favor other interests instead. Although political incentives come in many forms, citizens, organized advocacy groups, and competition serve as key avenues from which they originate. As previous research shows, both elected and appointed officials are responsive to these sources of pressure when deciding how to answer the pressing policy questions surrounding the environment (e.g., Potoski, 2002; Rabe, 2006; Daniels, Krosnick, Tichy, & Tompson, 2012; Ley & Weber, 2015). Consequently, some states may be willing to commit more resources, establish higher standards, or expect more in environmental quality, while others may only set out to do the bare minimum.

On the other hand, administrative capacities shape the practical realities of protecting the environment by creating limitations in what state agencies are capable of doing. As administrative agencies attempt to operationalize policies, they must find ways to marshal their human, financial, political, and infrastructure resources to accomplish policy goals. This tends to be easier said than done, and many administrative agencies face constraints which in turn limit their efficacy in policy implementation, program management, and environmental protection. Like political incentives, administrative capacities come in many forms. We focus on three dimensions: policymaking, managing information, and creating accountability. These dimensions help capture whether states are capable of making "good" policy choices, understanding environmental problems and solutions, and compelling both internal and external compliance, which tie to political science, public policy, and public administration perspectives on the conceptualization of capacities (Christensen & Gazley, 2008; Brinkerhoff & Morgan, 2010; Wu, Ramesh, & Howlett, 2015). In essence, these capacities are demonstrative of the broader ability of state environmental agencies to make, monitor, and enforce environmental policies. Subsequently, some states are efficient and effective in turning policy ideas into administrative reality and, in turn, serving as efficacious agents of environmental protection, while others flounder at every turn while trying to follow the same path.

Of course, these patterns of behavior have broader implications than just how policies are implemented within the 50 states. More specifically, how states choose to respond to the political and administrative challenges of protecting the environment sends shockwaves through the federal system as national, state, and local governments compete, collaborate, fight, and negotiate with each other in the process of achieving a collective mission to protect the environment. States are a worthy focal point in trying to understand this, as they serve as the linchpins of the federal system. From that middle ground, how states occupy policy arenas affects how national governments above and local governments below respond to environmental problems. In general, we should expect that if all three levels of government engage in positive interactions and organize themselves into an efficient multi-level governance system with each playing their respective roles, environmental protection is an achievable goal that can stand up to any metrics by which we measure its qualities (i.e., effectiveness, equity). But decades of experience and research on environmental federalism tell us that this scenario is unlikely, and the intergovernmental system is driven by opportunism and political conflicts as much as shared goals.

Through our examination, we have identified four types of states, as well as associated patterns of behavior. First, progressive states have both the political will to commit to environmental protection and the administrative capacities to operationalize that will, so they typically emerge as national leaders in environmental policy. Progressive states tend to be a product of both high public concern for the environment and advocacy group activity on the one hand, and capacities for policymaking and creating accountability on the other. This ordinarily manifests through aggressive policies that exceed national standards and rigorous enforcement, which lead to better environmental outcomes. We identified 15 progressive states that are mostly concentrated in the West, although a few also exist in the Northeast. From their perch as national leaders, progressive states gravitate towards being the aggressors in conflicts with the EPA and other federal agencies as they push the national government to be more progressive. Certainly, this was the case when Massachusetts sued the EPA in an effort to force the national regulation. Additionally, progressive states often develop sophisticated state-level institutions that leave local governments as compliance managers, such as in California, where a series of regional control boards supply the local capacities and expertise that cities and counties would otherwise provide. Unsurprisingly, these trends are apparent in climate policy, given that progressive states had already begun adopting innovative climate policies and working to mitigate their GHG production during the 2000s.

Second, struggler states have similar political incentives as progressive states but lack the requisite administrative capacities to match goals with reality. While struggler states may adopt innovative policies, they tend to flounder when protecting the environment in practice, which is usually evident by their struggles to meet national environmental standards. Struggler states usually occur where there is stiff horizontal competition from surrounding states and state agencies with capacities to create accountability. We identified 12 struggler states across the Midwest, Northeast, and South/Southeast. Given that administrative capacity presents an obstacle, struggler states typically seek out opportunities to supplement their deficits through collaborative and cooperative relationships with both national and local governments. Struggler states are generally open to creating partnerships and pursuing positive interactions but are also willing to engage in conflict if other policy actors put up barriers. For instance, North Carolina was interested in cooperation as a means to improve air quality, but the EPA's refusal to alter its strategy in regards to upwind states left the state with no choice but to take legal action. Alternatively, Florida has seen great success with collaborative watershed management by working with local governments, which supply expertise and leverage social capital, to achieve shared policy goals. Again, we see these trends beginning to play out in climate policy, where struggler states are adopting new, innovative policies and experiencing some limited success.

Third, delayer states have the administrative capacities to protect the environment but lack the political will to do so. Consequently, these states tend to delay environmental action when possible, but when forced to act, do well in meeting externally mandated goals. Delayer states are most common where public concern for the environment and advocacy group activity are low, but information management capacities are high. We identified 11 struggler states, with the majority in the South/Southeast and others spread across the Midwest, Northeast, and West. While delayer states have the capacity to make programs work, there is little interest among state leaders, so a typical tactic is to avoid conflict with the national government by maintaining the status quo. The quiet conflict between Louisiana's elected officials and the EPA over the use of the CAA's "general duties" clause is a prime example of state political leaders pushing to uphold the status quo. West Virginia illustrates similar behavior in dealing with local governments, where the WVDEP approaches CWA implementation as an exercise in compliance management but also engages stakeholder groups in order to find politically feasible options for addressing environmental problems at the community-level. While delayer states have adopted far fewer climate policies than struggler states, GHG emissions concentrations are similar, and based on our analysis of pollution concentrations from legacy environmental programs, these states may be poised for more success under a national climate policy than current trends may suggest.

Finally, regressive states lack both the administrative capacities and political incentives for environmental protection, so they generally put forth only minimal effort and experience a lower standard of environmental quality than the rest of the nation. Regressive states typically result from a lack of public concern for the environment and administrative capacities to create accountability. We identified 12 regressive states, mostly concentrated in the Midwest and South/Southeast, although one outlier exists in the Northeast. While they engage in conflict with the national government similar to progressive states, regressive states usually do it in an attempt to be openly defiant of pro-environmental agendas and/or to derail national programs. Texas's refusal to implement GHG permitting and the subsequent legal battle is a prototype for this behavior. However, regressive states also tend to put local governments in vulnerable positions as they are left holding the bag when states retreat from environmental protection, as illustrated by the struggles in Ohio that occurred when the state neglected to provide adequate support to local governments for maintaining their wastewater systems or collaboratively managing watersheds. Expectedly, regressive states experience both the highest concentrations of pollution and GHG emissions in the US and have taken little action to address climate change.

In sum, there are fairly predictable patterns of state behaviors related to environmental policies, so we should assume that the same pros and cons for programs like the CAA, CWA, and RCRA would reemerge with a national climate policy. On one hand, decentralization allows decisions to be made as close as possible to the people who are affected by those decisions, and in balancing the competing interests surrounding environmental and economic policy choices, it is easier to find consensus when choices are confined to narrow geographic areas (Woods & Potoski, 2010). Of course, this allows states to match policy strategies developed at the national-level with the socioeconomic, political, and technical challenges at the community-level. Furthermore, it also means that managers are positioned to establish operational localism by building stakeholder coalitions that legitimize administrative decisions (Reed, 2014). On the other hand, decentralization tends to make institutions more vulnerable to external stress, less resilient, and leads to inequitable distributions of outcomes (Meier & O'Toole, 2009). For instance, states may defect from national goals when economic or political pressures make it difficult to continue cooperating. Certainly, we have seen this with the "race-to-the-bottom," as states facing competitive pressures from their neighbors move towards lax regulatory regimes in order to attract industries (Potoski, 2002). By extension, we would have to expect significant variations across states in terms of both policies and outcomes, as states respond to both their political environment and the administrative realities of implementation.

But climate change is much different from legacy environmental problems that we have previously faced. Most importantly, the success of mitigation efforts will be highly dependent on cumulative GHGs at the national- and international-levels, as the effects of climate change cannot be disaggregated to impact only those states or communities that refuse to pull their own weight. That is, while ineffective implementation of the CAA may result in poor air quality in one state that presents only minimal risk to other states, the risks of ineffectively implementing a national policy to mitigate climate change are likely to be distributed around the nation and the globe. On the other hand, this could work to the advantage of adaptation efforts, where the benefits of doing so would be localized and the costs would be less likely to cross state lines. In other words, decentralization creates a substantial risk for success of a national policy if the focus is mitigation, but that risk is significantly reduced if the focus is adaptation. Further, the nature of climate change also breeds ambiguity surrounding the causes and effects, and, for policy, this means it is much more difficult to identify effective experiments and best practices, as well as establish political consensus. In comparison, we are confident in the environmental and economic impacts of the CAA because the science and the politics of air quality lend themselves to less ambiguity (Fowler, 2014).

As such, we may be able to follow a similar blueprint for a national climate policy that we previously used for legacy environmental programs, but we will also have to adapt as this new challenge highlights the inherent shortcomings of existing institutions. Under the most likely scenario that includes both mitigation and adaptation efforts, where the average state meets or nearly meets national standards, we should still have high expectations for success. Our experiences with the CAA, CWA, and RCRA would generally suggest that while a group of states will inevitably fall well short of any national standard, there are other states that will go above and beyond. If trends from pollutant concentrations and GHG emissions hold up (see Chapter 8), it is reasonable for us to expect that progressive states would balance out regressive states. However, this assumes that these asymmetries are taken into consideration early on in policy design, that progressive states are willing to bear more than their fair share of the burden, and that delayer and/or struggler states are not significantly deficient in compliance. Of course, this is contingent on finding a political consensus for what a national climate policy would entail; although recent history may give us reason to be circumspect in this regard, the political tides may be shifting (Wallach, 2019). Finally, this also assumes that mitigation efforts will not be too late to have any real impact.

In any regard, what we can conclude is that the impacts of a national climate policy will be largely dependent on the action or inaction of state governments. States are key players in environmental protection in the US, and much of the success made in improving air, water, and land quality since the 1970s is because of efforts at the state-level. Subsequently, it is highly unlikely that states would be willing to give up the policy space they have successfully occupied over the last half century, or that federal agencies would be able to replace the implementation expertise or capacities that they bring to the table. Furthermore, as previous chapters in this volume have explained, the efforts of national and local governments are at least partially in response to state action (or inaction), which positions states to not only directly provide environmental protection, but also to shape how environmental protection is provided as a whole in the US. For instance, even in the absence of a national climate policy, states have adopted and implemented their own policies, and almost half of states joined the US Climate Alliance following US withdrawal from the Paris Agreement (US Climate Alliance, 2019). In sum, overlooking environmental federalism as a component of climate policy is overlooking a key dimension in how environmental protection in the US functions.

Prospects for Environmental Federalism

While environmental federalism is a relatively niche area of scholarship, it remains essential as both an institutional framework for protecting the environment in the US and theoretically for understanding the federal system as a multi-level governance institution. As explained earlier, the shared power between national and state governments is not only an institutional legacy that stretches back to the founding of America and is unlikely to change, but it is also a core tenet of how environmental protection is provided. Of course, this presents pros and cons. On one hand, relying on federal institutions leads to collective action dilemmas, as cooperation and collaboration between governmental units runs into institutional barriers (Feiock & Scholz, 2009). Most importantly, federalism incentivizes state officials to take a myopic viewpoint in calculating the costs and benefits of policy action and largely limits their consideration of externalities that may manifest in other geographic areas. In many cases, working horizontally with other states and/or vertically with national or local governments is likely to result in more efficient, effective policy outcomes. Instead, we see instances in which states are more likely to eschew cooperation, as goal conflicts and uncertainty create risks and competition leads to political rewards (Wood & Bohte, 2004; Feiock & Scholz, 2009). As environmental problems rarely conform to the institutional or geographic boundaries of government, they regularly challenge whether federalist institutions are an administratively efficient approach to finding solutions.

Still, distributing authorities vertically between national, state, and local governments tends to be the most politically expedient approach, despite the constraints faced in some states (Volden, 2005). Above all else, it allows governmental units to experiment with politically feasible options for protecting the environment across a nation that is geographically, economically, and sociopolitically diverse. In large part, this hinges on states and their ability to adapt national programs to meet the unique needs of local communities and thereby solving one of the most pressing concerns of federal government: balancing the national versus the local. In essence, this permits environmental policy to satisfy the competing demands of one nation by allowing communities to determine their own best interest but also ensuring that those interests do not conflict with the greater good of society as a whole. Although both the political and administrative dimensions are vital to understanding environmental protection, our system of government tends to prioritize political expedience over administrative efficiency (Kettl, 2000; Bertelli & Lynn, 2006). Thus, as an institutional feature, it is unlikely that we can ever move away from federalism without changing the very foundations of American government.

As such, environmental federalism as a theoretical framework remains as important today as it was in the 1980s when scholars began employing institutional, intergovernmental relations, and federalism theory to understand how environmental policies were made and managed in the US (e.g., Bowman, 1985; Downey, 1985; Lester, 1986). At their core, theories of environmental federalism seek to understand how sovereign power is wielded over the environment and how we connect collective efforts to manage environmental problems to the institutions of our government. Given the fragmented regulatory regimes created by state lines, the inherent conflicts that emerge when different governmental units operate within the same policy space, and the changing environmental politics over the last several decades, these questions have not faded in magnitude of importance, even though environmental problems and policies as well as our understanding of them have changed significantly. In this volume, we have specifically contributed to this by examining the political and administrative challenges of environmental policy and the character of inter-actions between national, state, and local governments within the context of legacy environmental programs. While new and interesting questions are being presented on a continual basis as we consider and reconsider the government's role in mitigating and adapting to climate change, environmental federalism offers a useful lens for finding answers.

However, a key challenge is integrating this perspective with theories of environmental governance that take a broader view of environmental institu-tions by considering more holistically how both governmental units and NGOs collectively respond to environmental problems. From this perspective, gov-ernment is not the only means by which environmental problems are solved and policy actors self-organize around shared goals; and while national, state, or local governments may have a seat at the table, they do not wield hierarchical or coercive power over the other actors (Durant, Fiorino, & O'Leary, 2017). With the emergence of environmental NGOs and policy networks, scholarly atten-tion has shifted away from a predominant focus on what national or subnational governments are doing and towards a focus on how institutions function when nontraditional policy actors are involved and organization principles no longer conform to hierarchical or market-based models. Nevertheless, environmental governance and environmental federalism are interested in answering the same questions: who has the power to affect the environment, and how does that translate into collective action? Even though its role has been diminished in recent decades, there is no replacement for the sovereign power of government in addressing problems on the scale of air quality, watershed management, toxic waste disposal, or climate change. As such, theories of environmental federal-ism should not be entirely lost, as they play an important role in understanding the institutional legacies of American government in this policy arena.

Lingering Questions

Although our examination answered the questions that we initially posed, there are a few lingering questions that are worthy of mention here in as far as they guide additional inquiry into environmental federalism. First, are there other political or administrative factors that may help explain patterns of state behav-ior? Conceptually, both political incentives and administrative capacities can be broadly defined to incorporate numerous nuances that differentiate between states. While our measures proved rigorous in explaining patterns of pollu-tion and intergovernmental interactions as intended, we deliberately limited the number of variables in the interest of parsimony. Nevertheless, there is a litany of variables that previous scholars have used in analyzing this (or related) topics, such as political culture (e.g., Travis, Morris, & Morris, 2004; Hoornbeek, 2005), partisanship (e.g., Woods, 2008; Konisky & Woods, 2010), environmental expenditures (e.g., Bacot & Dawes, 1997), organizational structures (e.g., Ringquist, 1993; Fowler, 2013), and administrative design (e.g., Potoski & Woods, 2001). Considering further ways to measure our core concepts can potentially provide additional insights into the structure of the political and administrative dimensions of environmental policy within the federal system. This may be particularly important as it relates to differentiating between types of states and understanding the mechanisms to drive interactions with national and local governments.

Second, do the political and administrative dimensions also explain patterns of state behavior as it relates to horizontal intergovernmental relations and/or interactions with NGOs? A key tenet of our theory is that how states function as intermediaries in the federal system sends ripples vertically that affect both national and local governments, and we must also consider whether these ripples move horizontally. Given the tendency of environmental problems to transect state lines, states regularly engage in horizontal interactions. Like vertical interactions, when these are positive and cooperative, effective solutions are agreed to and implemented, but when conflict takes over, environmental problems linger and intergovernmental battles become unproductive (Bowman, 2004; Bowman & Woods, 2007; Schlager & Heikkila, 2009). Furthermore, as governments in general begin to rely more and more on NGOs to provide public services, how states engage in environmental protection may be important in dictating the terms of NGO involvement, which may range from contracting out to partnerships to independent efforts designed to compensate for deficient public programs (Kettl, 2006). Notably, within this horizontal dimension, policy actors lack hierarchical authority to force their interests on others, which likely leads to a more complex set of interactions. Although our focus is on the vertical dimension of federalism, the horizontal dimension is equally as important for understanding the broader frameworks of both environmental federalism and governance.

Third, how much do states change over time? As our analysis only focuses on the 2000s, we must consider whether our findings are generalizable outside this time period, and by extension, how stable the structure of political incentives and administrative capacities are. For instance, previous research suggests that public opinion is relatively stable over short periods of time, but changes do manifest over decades (Daniels, Krosnick, Tichy, & Tompson, 2012; Kim & Urpelainen, 2018). Additionally, states have partaken in administrative reforms at various rates since the 2000s, and some organizations have proven to be more capable of learning than others (Moynihan, 2005, 2006). Essentially, we may find that for certain states the political or administrative contexts have evolved. As a whole, this would suggest that we should be cautious in extrapolating these findings beyond the 2000s, as it is likely that our classifications of some states would be inaccurate during other decades. Nonetheless, applying this framework to other periods of time may be probative in understanding the evolution of state politics and administration as well as environmental federalism. This may be particularly important in regards to the progression of national-state conflicts or the emergence of local governments as policy innovators. In other words, they may be responding to changing positions of states as political incentives and/or administrative capacities become stronger or weaker.

Fourth, does any of this apply outside of the US? While our focus has been exclusively on American federalism, the US does not have a monopoly on this form of government. Other nations, such as Germany, Russia, and Brazil, are organized on the same principles of shared powers and decentralization. Additionally, many international organizations, such as the European Union, operate under similar structures, although they may not self-identify as federalist institutions. In many cases, their interest in dividing powers between national and subnational governmental units is in response to the same challenges that the US faces in terms of trying to unite diverse communities under a single banner or trying to organize collective action that is conscious of the local experience. Furthermore, federalism "can be understood as a way of approaching politics that acknowledges group identify alongside individual identity. However, it is a particular form of group identity that federalism acknowledges – a spatial, locational, or territorial one" (Hueglin & Fenna, 2015, p. 16). Certainly, as populations around the world become more politically active and conscious of the impeding environmental transitions on the horizon that will be location-based, there will be growing interest in balancing the group versus the individual consequences of responding to climate change.

As such, my findings may have implications for understanding environmental policy and governance in other countries as well. Although our specific indicators may not cross international borders well, our theory should be widely applicable. Specifically, the core tenets of our theory are not exclusive to the institutional structures, policies, political incentives, or administrative capacities in the US. Rather, our theory is based on the broader assertion that subnational government responds to a unique set of political circumstances and administrative constraints that are bound by jurisdictional boundaries in providing environmental protection. In large part, this is consistent with extant scholarship on comparative federalism (Burgess, 2006; Menon & Schain, 2006; Hueglin & Fenna, 2015). Furthermore, even in nonfederal nations or international organizations that are not as hierarchically organized as conventional federalist institutions, decentralization is an administrative and political reality to managing geographically dispersed, diverse populations. To this end, our findings and theory tell us as much about the pitfalls and plusses of decentralization in general as they do about federalism specifically. Consequently, examining this theory in the comparative context of other nations may provide further insights into federalist institutions and the utility of decentralization as a mechanism for coordinating government action.

Finally, if state behavior is indeed driven by these two dimensions, how can we operationalize this knowledge so that it leads to better environmental choices? In order to effectively respond to emerging environmental challenges, we must bridge the gap between research and practice (Bushouse et al., 2011). As such, it behooves us to think about how these findings may be applied in the real-world. Given that our goal here has been to understand environmental federalism in regards to legacy programs and to determine whether that understanding can be extrapolated to climate change, the implications of my work are not limited to students or scholars. To this end, some of my findings could be used to help guide strategic thinking about environmental policy. For instance, my findings could inform policymakers working to update the CAA, CWA, or RCRA, or to design a national climate policy, or environmental managers looking for a deeper understanding of the context in which their programs operate. Subsequently, we must also consider new research questions that seek to unpack environmental federalism in a way that improves environmental policy choices at national- and subnational-levels. In sum, environmental federalism is concerned with the core of how the environment is protected, and understanding it carries the potential to affect positive change.

References

Bacot, A.H. & R.A. Dawes. 1997. State Expenditures and Policy Outcomes in Environmental Program Management. *Policy Studies Journal* 25(3): 355–370.

Bertelli, A.M. & L.E. Lynn. 2006. *Madison's Managers: Public Administration and the Constitution.* Baltimore, MD: John Hopkins University Press.

Bowman, A.O'M. 1985. Hazardous Waste Management: An Emerging Policy Area Within an Emerging Federalism. *Publius* 15(1): 131–144.

Bowman, A.O'M. 2004. Horizontal Federalism: Exploring Interstate Interactions. *Journal of Public Administration Research & Theory* 14(4): 535–546.

Bowman, A.O'M. & N.D. Woods. 2007. Strength in Numbers: Why States Join Interstate Compacts. *State Politics & Policy Quarterly* 7(4): 347–368.

Brinkerhoff, D.W. & P.J. Morgan. 2010. Capacity and Capacity Development: Coping with Complexity. *Public Administration & Development* 30(1): 2–10.

Burgess, M. 2006. *Comparative Federalism: Theory and Practice.* New York, NY: Routledge.

Bushouse, B.K., W.S. Jacobson, K.T. Lambright, J.J. Llorens, R.S. Morse, & O. Poocharoen. 2011. Crossing the Divide: Building Bridges between Public Administration Practitioners and Scholars. *Journal of Public Administration Research & Theory* 21(i1): 99–112.

Christensen, R.K. & B. Gazley. 2008. Capacity for Public Administration: Analysis of Meaning and Measurement. *Public Administration & Development* 28(4): 265–279.

Daniels, D.P., J.A. Krosnick, M.P. Tichy, & T. Tompson. 2012. Public Opinion on Environmental Policy in the United States. In *The Oxford Handbook of U.S. Environmental Policy*, edited by M.E. Kraft & S. Kamieniecki (pgs. 461–486). Oxford: Oxford University Press.

Downey, G.L. 1985. Federalism and Nuclear Waste Disposal: The Struggle Over Shared Decision Making. *Journal of Policy Analysis & Management* 5(1): 73–99.

Durant, R.F., D.J. Fiorino, & R. O'Leary. 2017. *Environmental Governance Reconsidered: Challenges, Choices, and Opportunities*, 2nd ed. Cambridge, MA: MIT Press.

Feiock, R.C. & J.T. Scholz. 2009. Self-Organizing Governance of Institutional Collective Action Dilemmas: An Overview. In *Self-Organizing Federalism: Collective Mechanisms to Mitigate Institutional Collective Action Dilemmas*, edited by R.C. Feiock & J.T. Scholz (pgs. 3–32). Cambridge: Cambridge University Press.

Fowler, L. 2013. Measuring Organization: Performance in Environmental Agencies. *International Journal of Organizational Theory & Behavior* 16(3): 326–359.

Fowler, L. 2014. Assessing the Framework of Policy Outcomes: The Case of the U.S. Clean Air Act and Clean Water Act. *Journal of Environmental Assessment Policy & Management* 16(4): 1–20.

Hoornbeek, J.A. 2005. The Promises and Pitfalls of Devolution: Water Pollution Policies in the American States. *Publius* 35(1): 87–114.

Hueglin, T.O. & A. Fenna. 2015. *Comparative Federalism: A Systematic Inquiry*, 2nd ed. Toronto, Canada: University of Toronto Press.

Kettl, D.F. 2000. Public Administration at the Millennium: The State of the Field. *Journal of Public Administration Research & Theory* 10(1): 7–34.

Kettl, D.F. 2006. Managing Boundaries in American Administration: The Collaboration Imperative. *Public Administration Review* 66(s1): 10–19.

Kim, S.E. & J. Urpelainen. 2018. Environmental Public Opinion in the U.S. States, 1973–2012. *Environmental Politics* 27(1): 89–114.

Konisky, D.M. & N.D. Woods. 2010. Exporting Air Pollution? Regulatory Enforcement and Environmental Free Riding in the United States. *Political Research Quarterly* 63(4): 771–782.

Konisky, D.M. & N.D. Woods. 2016. Environmental Policy, Federalism, and the Obama Presidency. *Publius* 46(3): 366–391.

Konisky, D.M. & N.D. Woods. 2018. Environmental Federalism and the Trump Presidency: A Preliminary Assessment. *Publius* 48(3): 345–371.

Lester, J.P. 1986. New Federalism and Environmental Policy. *Publius* 16(1): 149–166.

Ley, A.J. & E.P. Weber. 2015. The Adaptive Venue Shopping Framework: How Emergent Group Choose Environmental Policymaking Venues. *Environmental Politics* 24(5): 703–722.

Meier, K.J. & L. O'Toole. 2009. The Proverbs of New Public Management: Lessons from an Evidence-based Research Agenda. *American Review of Public Administration* 39(1): 4–22.

Menon, A. & M. Schain. 2006. *Comparative Federalism: The European Union and the United States in Comparative Perspective*. Oxford: Oxford University Press.

Moynihan, D.P. 2005. Why and How Do State Governments Adopt and Implement "Managing for Results" Reforms? *Journal of Public Administration Research & Theory* 15(2): 219–243.

Moynihan, D.P. 2006. Managing for Results in State Government: Evaluating a Decade of Reform. *Public Administration Review* 66(1): 77–89.

Potoski, M. 2002. Clean Air Federalism: Do States Race to the Bottom? *Public Administration Review* 61(3): 335–343.

Potoski, M. & N.D. Woods. 2001. Designing State Clean Air Agencies: Administrative Procedures and Bureaucratic Autonomy. *Journal of Public Administration Research & Theory* 11(2): 203–222.

Rabe, B. 2006. Race to the Top: The Expanding Role of U.S. State Renewable Portfolio Standards. *Sustainable Development Law & Policy* 8(3): 10–17.

Reed, D.S. 2014. *Building the Federal Schoolhouse: Localism and the American Education State.* New York: Oxford University Press.

Ringquist, E.J. 1993. *Environmental Protection at the State Level.* Armonk, NY: ME Sharpe.

Schlager, E. & T. Heikkila. 2009. Resolving Water Conflicts: A Comparative Analysis of Interstate River Compacts. *Policy Studies Journal* 37(3): 367–392.

Travis, R., J.C. Morris, & E.D. Morris. 2004. State Implementation of Federal Environmental Policy: Explaining Leverage in the Clean Water State Revolving Fund. *Policy Studies Journal* 32(3): 461–480.

U.S. Climate Alliance. 2019. *About Us* [online]. Available at www.usclimatealliance. org/about-us [Retrieved January 1, 2019].

Volden, C. 2005. Intergovernmental Political Competition in American Federalism. *American Journal of Political Science* 49(2): 327–342.

Wallach, P.A. 2019. *Where Does US Climate Policy Stand in 2019? Brooking Institute Series on Regulatory Process & Perspective* [online]. Available at www.brookings.edu/2019/ 03/22/where-does-u-s-climate-policy-stand-in-2019/ [Retrieved January 1, 2019].

Wood, B.D. & J. Bohte. 2004. Political Transaction Costs and the Politics of Administrative Design. *Journal of Politics* 66(1): 176–202.

Woods, N.D. 2008. The Policy Consequences of Political Corruption: Evidence from State Environmental Programs. *Social Science Quarterly* 89(1): 258–271.

Woods, N.D. & M. Potoski. 2010. Environmental Federalism Revisited: Second-Order Devolution in Air Quality Regulation. *Review of Policy Research* 27(6): 721–739.

Wu, X., M. Ramesh, & M. Howlett. 2015. Policy Capacity: A Conceptual Framework for Understanding Policy Competences and Capabilities. *Policy & Society* 34(3–4): 165–171.

INDEX